AF581775

Dynamics of the Global Savanna and Grassland Biomes

Dynamics of the Global Savanna and Grassland Biomes

Editors

Hannah Victoria Herrero
Jane Southworth

MDPI • Basel • Beijing • Wuhan • Barcelona • Belgrade • Manchester • Tokyo • Cluj • Tianjin

Editors

Hannah Victoria Herrero
University of Tennessee
USA

Jane Southworth
University of Florida
USA

Editorial Office
MDPI
St. Alban-Anlage 66
4052 Basel, Switzerland

This is a reprint of articles from the Special Issue published online in the open access journal *Applied Sciences* (ISSN 2076-3417) (available at: https://www.mdpi.com/journal/applsci/special_issues/global_savanna_grassland_biomes).

For citation purposes, cite each article independently as indicated on the article page online and as indicated below:

LastName, A.A.; LastName, B.B.; LastName, C.C. Article Title. *Journal Name* **Year**, *Volume Number*, Page Range.

ISBN 978-3-0365-0348-6 (Hbk)
ISBN 978-3-0365-0349-3 (PDF)

Cover image courtesy of Hannah Victoria Herrero.

Contents

About the Editors

Hannah Herrero is an Assistant Professor at the University of Tennessee, Department of Geography. She holds holds a B.Sc. in Geography with Environmental Geosciences, an M.Sc. in Geography, and a Ph.D. in the fields of Geography with Digital Geography and GIS from the University of Florida, USA. Her current research interests are focused on the study of human-environment interactions within the field of Land Change Science through the use of remote sensing technologies. All of her research is undertaken with highly interdisciplinary research teams, which involve both physical and social scientists. Her particular strengths lie in the remote sensing of vegetation dynamics; land use and land cover change; the implications of scale and scaling in remote sensing and modeling analyses; and people and parks. Her work of understanding environmental change has a particular concentration on conservation and protected areas, as well as savanna science. While the work is applicable globally, she has a strong regional focus on southern Africa.

Jane Southworth is a Full Professor and the current Department Chair at the University of Florida, Department of Geography. She holds a B.Sc. in Geography, from Leicester University in the UK, an M.Sc. in Geography, with a Specialization in Meteorology and Climatology from Indiana University, and a Ph.D. in Environmental Science, also from Indiana University. Professor Southworth's research interests are based on the study of social-ecological systems within the field of Land Change Science and Geospatial Science. Her particular foci are: remote sensing of vegetation dynamics with a focus on time-series approaches to remotely sensed analyses and linkages with climatic drivers; linking changes in land use and land cover change to their respective drivers with a focus on land change modeling; the implications of scale and scaling in remote sensing and modeling analyses; addressing linkages and drivers of change as they relate to people and parks; and modeling of the impacts of climate change and changing climate variability on human-environment systems and vegetation dynamics. Dr. Southworth has been part of over $18 million in funded interdisciplinary research projects, published 2 books, over 90 peer-reviewed journal articles, and has successfully advised 20 Ph.D. and 8 Masters students, with another 5 Ph.D. students currently in progress.

Preface to "Dynamics of the Global Savanna and Grassland Biomes"

This special issue includes five published papers with novel insights that range on specific topics from: modeling effects of climate change on productivity of rangelands in Zimbabwe, understanding the health of savanna vegetation in and around national parks in Southern Africa and Belize over time, a long-term field study assessing the effects of restoration efforts on degraded meadow steppes in northern China, and commentary on socio-environmental dynamics of alpine grasslands in western China.

Hannah Victoria Herrero, Jane Southworth

Editors

Editorial

Special Issue on Dynamics of the Global Savanna and Grasslands Biomes

Hannah Victoria Herrero [1,*] and Jane Southworth [2]

1 Department of Geography, University of Tennessee, 1000 Philip Fulmer Way, Room 315, Knoxville, TN 37996-0925, USA
2 Department of Geography, University of Florida, 3141 Turlington Hall, Gainesville, FL 32611, USA; jsouthwo@ufl.edu
* Correspondence: hherrero@utk.edu; Tel.: +1-865-974-6043

Received: 9 October 2020; Accepted: 11 November 2020; Published: 13 November 2020

1. Summation

Savanna and grassland biomes cover more of the earth's surface than any other biome type, and yet they are still largely understudied. In recent decades, global savanna and grassland ecosystems have become more prominent in the literature focused on global change dynamics. Savanna and grasslands represent unique biomes with their own challenges, both in terms of their study and in terms of their complexity, leading to many contradictory and often controversial findings. The global threats to these systems are potentially significant—from climate change impacts to human management challenges, from possible degradation to complete desertification, and looking across varied disturbance regime shifts.

This Special Issue of *Applied Sciences*, "Dynamics of Global Savanna and Grassland Biomes", is intended for a wide and interdisciplinary audience, and covers recent advances around the themes of drivers of vegetation dynamics, further understanding carbon interactions in these critical landscapes, advances in modeling both current and future system states, tipping points in savanna systems, human-environment interactions and challenges for management, and biodiversity and ecosystem services.

This Special Issue includes five published papers with novel insights that span a number of specific topics: modeling effects of climate change on the productivity of rangelands in Zimbabwe [1], understanding the health of savanna vegetation in and around national parks in Southern Africa and Belize over time [2,3], a long-term field study assessing the effects of restoration efforts on degraded meadow steppes in northern China [4], and commentary on socio-environmental dynamics of alpine grasslands in western China [5].

Acknowledgments: This issue would not be possible without the contributions of various talented authors, hardworking and professional reviewers, and dedicated editorial team of *Applied Sciences*. Congratulations to all authors—no matter what the final decisions of the submitted manuscripts were, the feedback, comments, and suggestions from the reviewers and editors helped the authors to improve their papers. We would like to take this opportunity to record our sincere gratefulness to all reviewers. Finally, we place on record our gratitude to the editorial team of *Applied Sciences*, and special thanks to Carla Gao.

Conflicts of Interest: The authors declare no conflict of interest.

References

1. Senda, T.S.; Kiker, G.A.; Masikati, P.; Chirima, A.; van Niekerk, J. Modeling Climate Change Impacts on Rangeland Productivity and Livestock Population Dynamics in Nkayi District, Zimbabwe. *Appl. Sci.* **2020**, *10*, 2330. [CrossRef]

2. Herrero, H.; Southworth, J.; Muir, C.; Khatami, R.; Bunting, E.; Child, B. An Evaluation of Vegetation Health in and around Southern African National Parks during the 21st Century (2000–2016). *Appl. Sci.* **2020**, *10*, 2366. [CrossRef]
3. Blentlinger, L.; Herrero, H.V. A Tale of Grass and Trees: Characterizing Vegetation Change in Payne's Creek National Park, Belize from 1975 to 2019. *Appl. Sci.* **2020**, *10*, 4356. [CrossRef]
4. Xu, L.; Nie, Y.; Chen, B.; Xin, X.; Yang, G.; Xu, D.; Ye, L. Effects of Fence Enclosure on Vegetation Community Characteristics and Productivity of a Degraded Temperate Meadow Steppe in Northern China. *Appl. Sci.* **2020**, *10*, 2952. [CrossRef]
5. Feng, H.; Squires, V.R. Socio-Environmental Dynamics of Alpine Grasslands, Steppes and Meadows of the Qinghai–Tibetan Plateau, China: A Commentary. *Appl. Sci.* **2020**, *10*, 6488. [CrossRef]

Article

Modeling Climate Change Impacts on Rangeland Productivity and Livestock Population Dynamics in Nkayi District, Zimbabwe

Trinity S. Senda [1,*], Gregory A. Kiker [2,3,*], Patricia Masikati [4], Albert Chirima [5] and Johan van Niekerk [6]

1 Department of Land Resource Management and Agricultural Technology, University of Nairobi, Nairobi 00625, Kenya
2 Department of Agricultural and Biological Engineering, University of Florida, Gainesville, FL 32611, USA
3 School of Mathematics, Statistics and Computer Science, University of KwaZulu-Natal, Pietermaritzburg 3209, South Africa
4 World Agroforestry Centre (ICRAF), Lusaka, Zambia; P.Masikati@cgiar.org
5 National University of Science & Technology, PO Box AC 939, Ascot, Bulawayo, Zimbabwe; albert.chirima@nust.ac.zw
6 Centre for Sustainable Agriculture, University of the Freestate, Bloemfontein 9300, South Africa; vniekerkJA@ufs.ac.za
* Correspondence: tssenda@students.uonbi.ac.ke (T.S.S.); gkiker@ufl.edu (G.A.K.)

Received: 26 February 2020; Accepted: 22 March 2020; Published: 28 March 2020

Abstract: Smallholder farmers in semi-arid areas depend on both cropping and livestock as the main sources of livelihoods. Rangeland productivity varies on both spatial and temporal scales and provides the major source of feed for livestock. Rangeland productivity is expected to decline with climate change thereby reducing livestock feed availability and consequently livelihoods that depend on livestock. This study was carried out to assess the impacts of climate change on rangeland productivity and consequently livestock population dynamics using a 30-year simulation modeling approach. The climate scenarios used in the simulations are built from the localized predictions by General Circulation Models (GCMs). The primary climate variables under consideration are rainfall (+/−7% change), carbon dioxide (CO_2 up to 650 ppm) and temperature (+4 °C change). This was done by applying the SAVANNA ecosystem model which simulates rangeland processes and demographic responses of herbivores on a temporal and spatial scale using a weekly internal time step and monthly spatial and temporal outputs. The results show that rainfall levels of less than 600 mm/year have the largest negative effect on herbaceous biomass production. The amount of biomass from the woody layer does not change much during the year. The carbon dioxide (CO_2) effects are more influential on the tree and shrub layers (C_3 plants) than the herbaceous layer (C_4 grasses). The CO_2 effect was more dominant than the effects of rainfall and temperature. In the baseline simulations, the shrub plant layer increased significantly over 30 years while there is a three-fold increase in the woody plant layer (trees and shrubs) where biomass increased from a 1980 production to that of 2010. The biomass of the herbaceous layer was stable over the historical period (1980 to 2010) with values fluctuating between 200 and 400 g/m^2. Grass green biomass has a variable distribution where most production occurred in the fields and cleared areas while lower levels of production were found in the forested areas. The spatial distribution of shrub green biomass was less directly linked to yearly rainfall. Shrub biomass was mostly found in forested areas, and it showed a steady increase in production. Cattle, donkey, and goat populations rose slowly from 1980 but the rise was disrupted by a dry period during the late 1980s to the early 1990s causing a decline in all populations primarily due to grass unavailability. The populations of cattle goats and donkeys started to rise again from 1995 onwards due to improvements in rainfall. Cattle and donkey populations were rising faster than that of goats while sheep population was not changing much for most of the simulation period, otherwise they declined significantly during the drought of 2002. Similar changes in simulated grass biomass

(g/m^2) were observed in almost all climate scenarios, except for the peak and low years. The livestock population simulation showed few variations in livestock population under all scenarios. The main conclusion from the study is that CO_2 effects on rangeland productivity are much more dominant than the localized effects of rainfall and temperature. This has implications of favoring the growth of the tree and shrub layers over herbaceous layer, which meant that in the long run, the species that are able to use tree and shrub layers may be kept as a livelihood source as they will have a feed source.

Keywords: climate change; SAVANNA; simulation; scenarios; livestock; rangeland productivity; CO_2 effects

---

1. Introduction

Climate change and variability are some of the most influential factors that impact the livelihoods of livestock farmers in semi-arid areas [1]. This is mainly through their effects on rangeland productivity which plays a critical role as the main feed source for livestock. [2], predicted a 25% loss of livestock production in crop-livestock systems in developing countries as a result of climate change mainly because livestock take time to rebuild after die-offs. In turn, livestock also plays an important role in the livelihoods of farmers by contributing significantly toward food security by alleviating seasonal food shortages through meat and milk and most importantly cash income [3]. However, high mortalities and low productivity were to be the most important constraints in this sector, hence farmers fail to realize the potential benefits from livestock. Poor access to animal health support and technologies as well as dry season feeding shortages contribute immediately to high mortalities [4].

The Intergovernmental Panel on Climate Change reported that the average temperature in Sub Saharan Africa (SSA) is projected to increase between 1.5 °C and 3.1 °C by 2050 [5]. It is also noted that extreme weather events such as droughts, floods, and changes in the frequency and intensity of dry spells are increasing in Southern Africa. The upward trend in temperature is projected to continue beyond 2050 while trends for precipitation vary spatially and from one model to another. The exact nature and extent of the impact of climate change on temperature and precipitation distribution pattern remain uncertain and it is expected to affect the poor and vulnerable societies especially in SSA. Despite the potential to increase agricultural productivity in SSA, climate change remains the greatest challenge. Africa is the most vulnerable region to climate change because widespread poverty limits the people's adaptive capacity [6]. The impacts of climate change on agriculture could seriously worsen the livelihood conditions for the rural poor and increase food insecurity in the region. Climate change is however a gradual process which can be adapted to depending on the resiliency of communities or countries affected.

Carbon dioxide levels in the atmosphere are expected to increase significantly in 2050 [5,7]. This will increase the efficiency of photosynthesis and water use, but also is expected to have greater impacts on C_3 plants. High temperatures tend to increase lignification of plant tissues and hence decrease the digestibility of forage. It is also predicted that climate change will induce a shift from C_3 to C_4 grasses. C_4 plants are more efficient in terms of photosynthesis and water use than C_3 plants. The C_3 forage plants generally have higher nutritive values, but lower yields, while C_4 plants contain large amounts of low-quality dry matter and have a higher carbon–nitrogen ratio [8].

Rangelands are under pressure from different drivers such as climate change, degradation, and human activities [9]. The productivity of rangelands in the semi-arid areas varies with time and space with most of them being characterized by grasses, shrubs, forbs, and browse [10]. Rangeland productivity and variability are closely dependent on several ecological conditions such as temperature, rainfall, and soil moisture content. These conditions are known to affect the species composition, abundance, and diversity. Different livestock species have different feeding behaviors and changes in rangeland condition and productivity may have far reaching consequences on grazing regimes [10–12].

These alterations may have adverse effects on livelihoods that depend on livestock for food and income. The spatial variability of the effects of climate change impacts within rangelands is likely to create inequitable distribution of feed resources [10].

The Malthusian theory of population suggests that the number of individuals (animals) are determined by the resources available creating an equilibrium situation. The concept of equilibrium and non-equilibrium in rangelands has been debated for many years [13]. Some authors argue that ungulate populations are more closely influenced by abiotic factors other than forage availability and density dependence. Boone and Wang [14], however, state that rainfall less than 300 to 400 mm per annum or an annual coefficient of variation of above 30% makes non-equilibrium dynamics more likely. In semi-arid areas, feed constraints are already being felt in terms of scarcity, fluctuating quantity and quality which may even degrade faster in communal areas.

Significant research has been conducted in terms of assessing the vulnerability and adaptation of the smallholder farmers to climate change [15,16]. There is however still much work to be done to improve the understanding of the implications that climate change may have in rangelands and how livestock farmers need to position themselves to hedge against the risks and shocks. There is limited information on rangeland productivity and the extent to which it changes spatially and temporally under different climatic conditions. The spatially explicit nature of rangelands in semi-arid areas together with the temporal variability in rainfall complicates the analysis and planning for sustainable livestock production and development. Because of the varying responses of different areas to climate change there is need for more localized assessments of climate change impacts on rangelands. Farmers need this information so that they can anticipate how the feed base will change and to what extent would it impact their livestock population dynamics under different climatic conditions, which would help them to be able to make informed decisions about sales, feeding and breeding. This paper conducts an initial assessment of the implications of climate change on rangeland condition in terms of feed availability (both quality and quantity) and its variability at both spatial and temporal scales.

This effort is achieved by applying the SAVANNA ecosystem model [17–19] under different climate scenarios within a southern African, communal rangeland landscape.

Objectives

- To analyse the trends in biomass (herbaceous, shrub and tree layers) production in relation to rainfall in Nkayi district (Zimbabwe) over a period of 30 years (1980–2010).
- To assess the effect of potential changes in climatic variables (rainfall, temperature, and atmospheric carbon dioxide) on rangeland productivity and livestock population dynamics on a temporal and spatial scale.
- To analyse how the biomass production from the tree, shrub and grass layers vary spatially in different times of the year.

2. Methodology

2.1. Study Area

The study was carried out in the communal areas of Nkayi district (19° 49′ 59″ S, 28°00′00″ E) of Matabeleland North Province in the central-western part of Zimbabwe (Figure 1). Nkayi district covered an area of 5320 km^2 which consist of communal areas and a section of protected forest (Gwampa Forest Reserve). The area has always had crop and livestock farming as the main land uses although there is an expansion of settlements. The agro-ecological conditions in Nkayi are characterized by unreliable rainfalls ranging between 450 and 650 mm/year, and periodic drought spells experienced during the rainy season [4]. The Nkayi district is divided into two zones that are mainly characterized by soil and vegetation types. The area in the southern part and along the Shangani River has the red sand loams also known as isibomvu/isidaka while the northern part toward Kana river consists mainly of Kalahari sands that are inherently infertile.

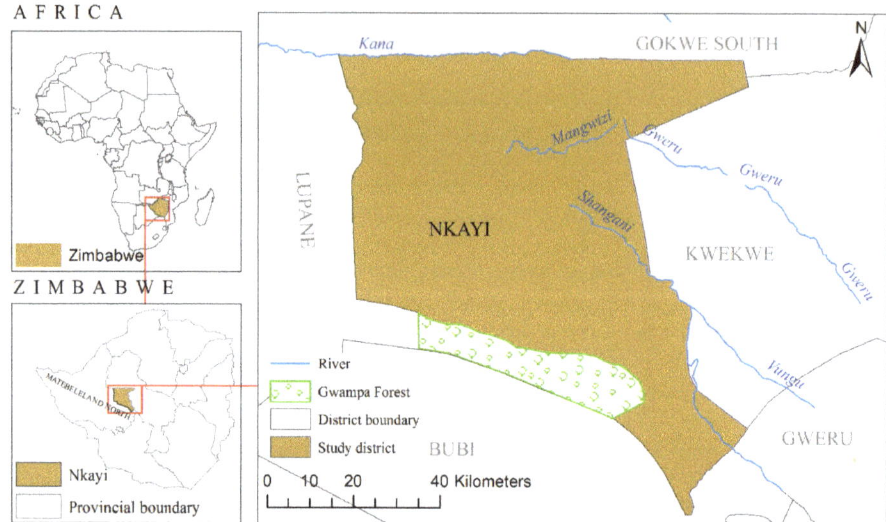

Figure 1. Location of Nkayi district in Matebeland north province of Zimbabwe (Data source: International Crop Research Institute for the Semi-Arid Tropics (ICRISAT)).

The vegetation in the red sandy loams is dominated by *Colophospermum mopane*, *Combretum apiculatum*, and various *Acacia* spp. The riverine strips are dominated by different tree species such as *Euclea divinorum*, *Acacia* spp, *Dichrostachys cinerea*, and *Albizia amara*. The northern and greater part of Nkayi District is called sand veld zone and is characterized by diverse miombo woodland dominated by *Julbernadia globiflora* and *Branchstegia Speciformis*. In areas that were previously cultivated regeneration is dominated by *Terminalia serecea* and *Julbernadia globiflora*. Human population densities apart from Nkayi business center ranges from 5–50 pp/km^2 with density decreasing from the South to North [3,4].

2.2. The SAVANNA Modeling System

This study applies the SAVANNA model (Figure 2) ([17,18,20–22], a grid-based, ecological modeling system. SAVANNA is a spatially explicit and process-oriented model that was developed to address localized management questions for different ecosystem types that include grasslands, shrublands and forests within the savanna biome. The early development and application of the SAVANNA model was in Turkana district of Kenya and improvements to the model were made in subsequent applications [23–25]. The model was designed to address the spatio-temporal variability of ecosystems by using remotely sensed data, GIS databases as well as spatial simulations in order to compute rates of plant production, forage intake by animals and ecosystem functions [26]. It simulates processes in an ecosystem using a weekly time step, simulating vegetation quantity and distribution in response to climate inputs as well as vegetation types consumed by herbivores [19]. The model also simulates the spatial redistribution of herbivores in response to changes in vegetation quantity as well as herbivore demographic responses to changes to biotic and abiotic factors. One advantage of the more meso-scale-focused SAVANNA model to more globalized, savanna ecosystem models such as G-Range [27], aDGVM [28] and LPJ-GUESS [29] is that localized conditions, impacts and potential mitigation alternatives can be simulated and explored in spatial and temporal detail that are meaningful to local managers and stakeholders.

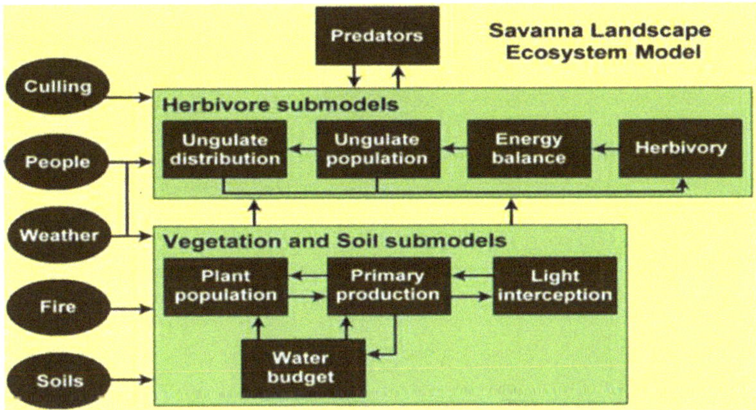

Figure 2. Schematic representation of the SAVANNA model [17].

2.3. Model Parameterization and Calibration

The SAVANNA model was parameterized to describe the response of changes of different plant and livestock groups. Weather data from 1980 to 2010 were used as the base scenario (Department of Meteorological Services, Zimbabwe) The model parameters used for this study are based on previous SAVANNA research [18,20–22,30] a goat production report by [4] and initial population data collected from Veterinary department in Nkayi [31].

Plants were placed into six functional groups based on previous modeling research [14,18,22,23] to simplify parameterization and to provide computational efficiency when executing simulations. The plant functional groups were based on functional groups described by [32] and classified as follows: Grass (C_4 photosynthetic pathway), shrubs (C_3 photosynthetic pathway), fine-leaved palatable trees, broad-leaved palatable trees, broad-leaved unpalatable trees and *Colophospermum mopane*. In addition, the grass layer was parameterized with the C_4 photosynthetic pathway while all woody shrubs and woody trees were parameterized as C_3 photosynthetic pathway species [33]. Livestock groups were also placed into functional groups for similar parameterization and computational efficiency. Herbivore functional groups consisted of cattle sheep, goats, and donkeys. Parameters for herbivore groups were provided by [18] and [34].

As input maps/grids, SAVANNA uses geographic layers (ESRI/ArcMap grids) describing elevation (United States Geological Survey [35] slope, aspect, vegetation, and land use [36], soil class [37], and distance to water to simulate the growth of plants and distribution of livestock. Each grid cell was divided horizontally into one of three vegetation types and vertically into layers of vegetation facets (grass-dominated, shrub-dominated, and tree-dominated) and soils (deep Kalahari sands, Brown loamy sands and Grayish Brown sands). As GIS grids are used as input into SAVANNA, the model outputs both time series files and GIS grids as well. In this study, a 2 km × 2 km grid cell size was used to model the rates of plant production, response of animals to forage availability and ecosystem function under varying climatic conditions. The 2 km grid resolution was selected to provide computational efficiency yet enough detail to simulate the varying vegetation and livestock areas over the entire communal areas of the Nkayi district. The soil input grid was creates using the three soil classes described by [37] (Figure 3) with soil-layer moisture and carbon characteristics derived from [38] and [39].

The vegetation and land use maps were derived from previous land degradation studies [36] and converted to 2 km grid resolution (Figure 4) along with estimates of woody cover and shrub/grass biomass for three primary vegetation types (forest, tilled/fallow field, and degraded). Initial grass biomass amounts were set low to reflect low forage production [36] with 25 g/m^2 for fields, 35 g/m^2 for conservation forest areas and 10 g/m^2 for degraded areas. Woody tree cover in forest vegetation areas was initialized at 70% (20% fine-leafed palatable, 20% broad-leafed palatable, 20% fine-leafed unpalatable and 10% mopane). Woody tree cover in degraded vegetation areas was initialized at 40% (10% fine-leafed palatable, 10% broad-leafed palatable, 10% fine-leafed unpalatable and 10% mopane). Woody tree cover in field areas was initialized at minimal levels (<5%) to represent some presence around human dwellings. Woody shrub biomass levels were initialized at 30 g/m^2 and 0.9 m in height for forest and degraded vegetation types and 5 g/m^2 and 0.1m for field areas to represent fallow zones and field borders.

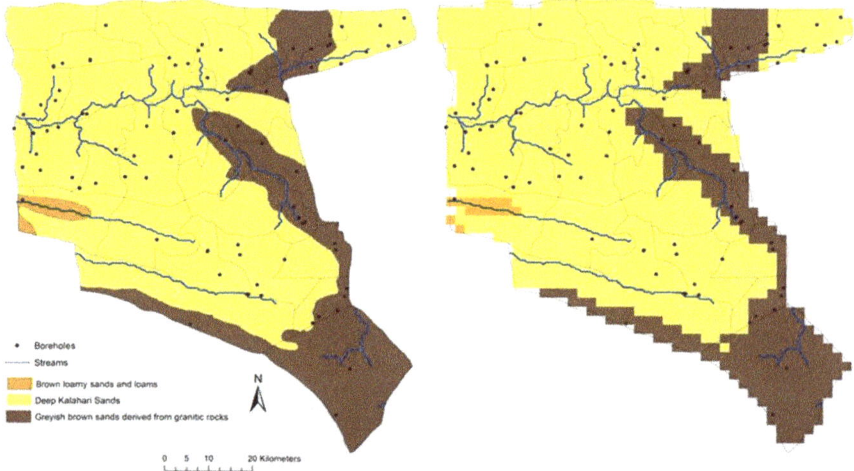

Figure 3. Nkayi soils map converted to 2 km resolution grid is used as inputs to the SAVANNA model.

Figure 4. Nkayi District land use map converted to 2 km resolution grid used as inputs to the SAVANNA model.

The elevation, slope and aspect were all obtained from the USGS Shuttle Radar Topography Mission (SRTM), 1-arc-second (90 m), digital elevation map [35], that was resampled to 2 km grid resolution (Figure 5). As rainfall in the region has a strong effect on seasonal browse production [10] and livestock performance, two water availability maps were created to represent dry and wet season ephemeral water availability for livestock along with perennially available boreholes. Seasonal water availability (Figure 6) was derived from combining distance to river (perennial and ephemeral) and borehole point data from [36] to achieve the minimum distance (m) for the average wet season (November to April) and the average dry season (May to October).

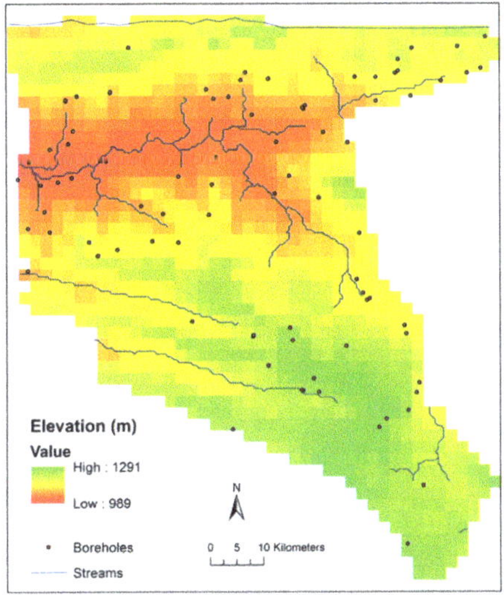

Figure 5. Digital elevation map (2 km grid resolution) for the Nkayi district. The grid was resampled from the worldwide SRTM 90 m elevation grid [35].

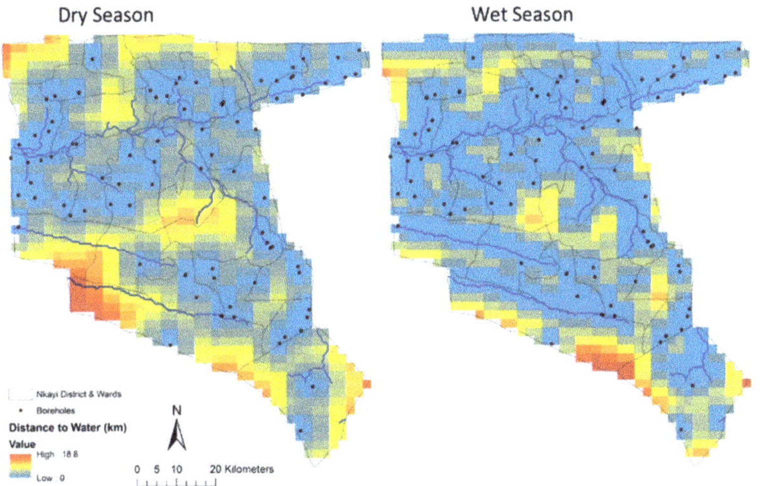

Figure 6. Distance to water (km) maps used as inputs to the SAVANNA model. The wet season grid was used in simulations from November to April while the dry season grid was used from May through October.

SAVANNA did not simulate any crop production in the agricultural areas but treated it as available forage for the season. Given that there was significant wildfire suppression and prevention efforts active in the region due to its proximity to forest conservation areas [40], the SAVANNA fire module was switched off (via parameter settings). Thus, any anthropogenic fires were assumed to be of limited size and low intensity as well as fully contained against any accidental spread. As a result, fire dynamics were not simulated in any of the climate scenarios. Additionally, shrub and tree biomass collection for firewood was not present in this version of the model.

Livestock populations in Nkayi district were set to levels described by the Department of Veterinary Services [31] with the total number of cattle at 110,000, one TLU being an animal weighing 350 kg. This level gives stocking rates of approx. 5.0 ha/TLU for cattle. Goat, sheep, and donkey populations stood at 25,000, 15,000, and 10,000, respectively [4]. SAVANNA simulations were executed with two sets of herbivore population demography options to explore the effect of herbivore population on rangeland productivity. The first option was to set all livestock population to constant values based on their initial 1980 levels. The second option allows livestock populations to rise and fall with effects from rangeland condition and water access. In both options, livestock populations were distributed according to habitat suitability with respect to forage, water, and animal densities. This meant that no specific field ownership of specific herds was assigned to control grazing locations as is the norm in most community grazing schemes, so the livestock were generally distributed according to the most satisfactory forage and water suitability. Additionally, no wildlife or other herbivores were simulated in addition to livestock.

2.4. Climate Change Scenarios

The Agriculture Model Inter-comparison and Improvement Project (AGMIP), assessed the 20 different GCMs and adopted five models namely CCSM4, GFDL-ESM2M, HadGEM2-ES, MIROC5 and MPIESM [41,42] These five adopted models were used as the basis of this study. These GCMs were chosen as they have a long history in development and evaluation, preference for higher resolution and established performance in monsoon regions ([42]. In the Nkayi district, rainfall was generally predicted to vary by +/−7% while for temperature they are predictions range up to 4 degrees Celsius above historic means. SAVANNA was used to simulate outcomes in the rangeland and livestock numbers from these scenarios. This was done by comparing the current/historic distribution, with simulated climate scenarios. In consultation with Agriculture Model Inter-comparison and Improvement Project (AGMIIP) partners, the different scenarios, and the core rangeland questions for analysis are presented in Table 1. For each climate scenario, the historical thirty-year dataset was altered to uniformly increase the maximum and minimum temperature by an additional 4 degrees Celsius, while rainfall was either increased or decreased by 7% on all monthly inputs. Additionally, each climate scenario was simulated with CO_2 levels set at 400 ppm and at 650 ppm to explore the effects of elevated CO_2 on the different plant layers. As a result, there were six different scenario simulations for analysis.

Table 1. Core questions that guide SAVANNA simulations with historic and GCM Scenarios.

<table>
<tr><th rowspan="2">Core Questions</th><th colspan="4">GCM Scenarios</th></tr>
<tr><th colspan="4">Historic Scenario (1980–2010)</th></tr>
<tr><td rowspan="5">1. What are the temporal and spatial dynamics of biomass production from the three layers? (tree, shrub, herbaceous)
2. How do the livestock populations respond to changes in the vegetation?</td><td>1.</td><td>No change in historical rainfall and temperature</td><td>2.</td><td>With/without increased carbon dioxide (400 ppm to 650 ppm)</td></tr>
<tr><td colspan="4">Hot and wet scenario</td></tr>
<tr><td>3.</td><td>7% increase in rainfall and 4 °C increase in temperature</td><td>4.</td><td>With/without increased carbon dioxide (400 ppm to 650 ppm)</td></tr>
<tr><td colspan="4">Hot and dry scenario</td></tr>
<tr><td>5.</td><td>7% decrease in rainfall and 4 °C increase in temperature</td><td>6.</td><td>With/without increased carbon dioxide (400 ppm to 650 ppm)</td></tr>
</table>

3. Results

The results below show the biomass production and livestock population dynamics in the historical and climate scenarios listed in Table 1. SAVANNA simulates biomass quantities at a weekly time step. The green biomass (g/m^2) for the herbaceous, shrub and tree layers are used to estimate the peak production period which was selected from the 3rd week of March every year of simulation. This section describes the historical simulation (1980–2010) results as a baseline scenario for subsequent comparison to the subsequent temperature/rainfall/CO_2 combinations.

3.1. Historical Scenario: Biomass Production over the Entire Nkayi District

The overall Nkayi district average trends in biomass production over the baseline (historical) period of 30 years (1980 to 2010) in relation to rainfall totals are presented in Figure 7.

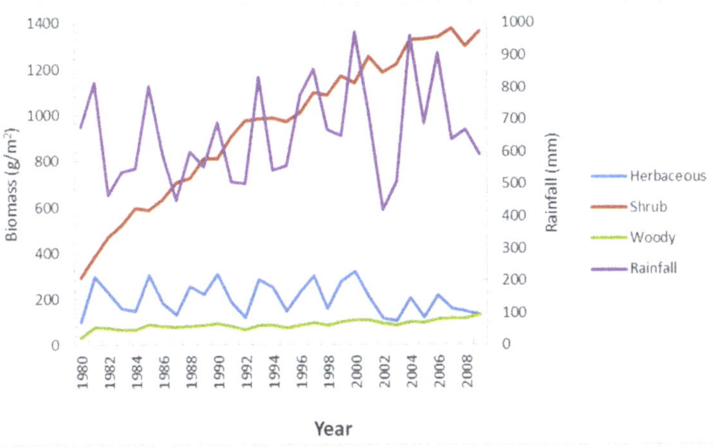

Figure 7. Trends in peak biomass production (g/m^2) over a period of 30 years (Historic scenario without climate change effects).

In these baseline simulations, the shrub layer increased significantly over 30 years from about 300 g/m^2 in the early 1980s to over 1200 g/m^2 in the year 2010. A three-fold increase was realized in the woody plant layer (shrubs and trees) where biomass increased from 1980 production levels of approximately 30 g/m^2 to being almost steady around approximately 100 g/m^2 by 2010. The herbaceous layer does not increase over the historical period with biomass fluctuates between 200 and 400 g/m^2 over the years, primarily in response to rainfall amounts. Rainfall levels of less than 600 mm/year have the largest negative effect on herbaceous biomass production. Specific drought years such as 1992 and 2002 have production levels close to 120 g/m^2. Conversely, rainfall totals over 1000 mm/year positively influenced biomass production levels to over 300 g/m^2. Both the shrub and woody layers have less direct connection to seasonal rainfall totals. A study carried out by [43] on modeling the semi-arid grazing systems showed a pattern similar to the findings of this study on the trends of biomass production. Using the average rainfall and the monthly averages of the biomass yields for the 30-year period a general pattern of the relationship between the rainfall and the biomass was established. The grass layer fluctuates around fluctuating rainfall while the tree and shrub layers are less sensitive to variations. The amount of biomass from the woody layer does not change much during the year. The herbaceous and the shrub layer follow the same pattern, but the shrubs are much higher than the herbaceous layer. The herbaceous layer peaks around February and March and is at its lowest during the August to October period. (Figure 8).

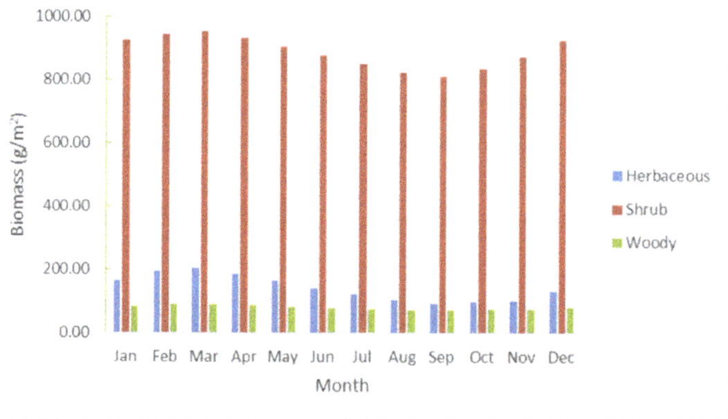

Figure 8. The monthly variations in biomass of the three layers during the year (averages for the 30-year period).

3.2. Historical Scenario: The Spatial Distribution of Green Biomass Production

Figures 9–11 show how the biomass from the three layers has been distributed in Nkayi over a period of 30 years in the historic scenario. Within Figure 9, grass green biomass has a variable distribution with most production occurring in the fields and cleared areas with chronically lower levels (2–50 g/m^2) than found within the forest areas. As was seen in earlier time series graphs, yearly production mostly follows rainfall. Within this simulation, SAVANNA does not simulate crop production areas for part of the seasons so that the availability of forage biomass is probably more limited in reality than shown in these maps in Figures 9–11.

Figure 10 shows the spatial distribution of shrub green biomass which is less directly linked to yearly rainfall. Shrub biomass is mostly found in forest areas, but a significant and steady increase in production is found in all spatial areas over the 30-year simulation. This systematic production increase is mostly likely due to the overall suppression of fire in the system and the exclusive grass consumption of most livestock except goats. In addition, SAVANNA does not simulate any domestic removal of shrub biomass for charcoal, cooking, or other uses.

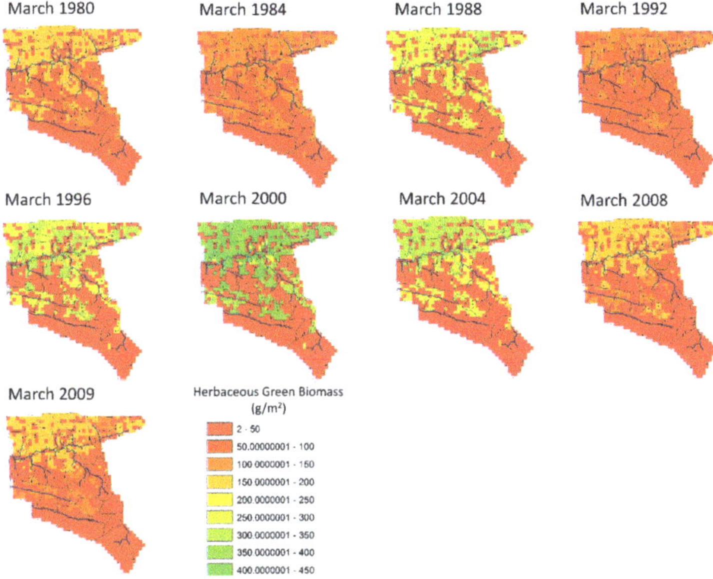

Figure 9. The spatial distribution of herbaceous green biomass for selected years (March every 4th year of the historic scenario).

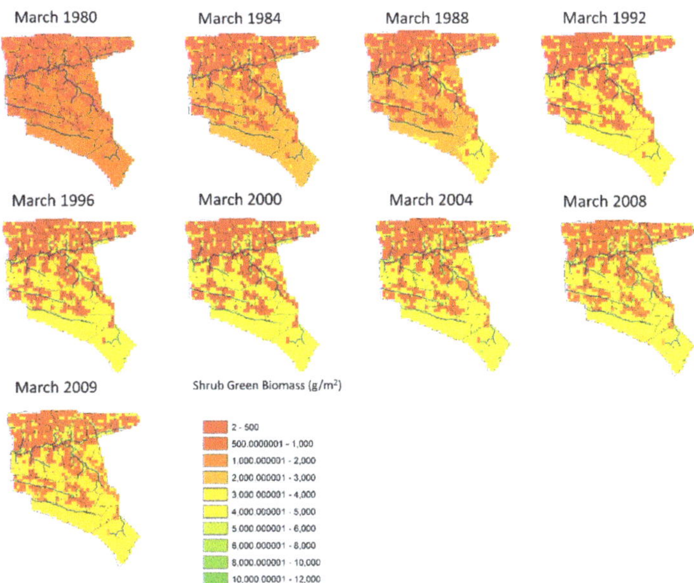

Figure 10. The spatial distribution of shrub green biomass for selected years (March every 4th year of the historic scenario).

Figure 11 shows the spatial distribution of tree green biomass which increases at a faster rate across all height classes. The tree green biomass does follow the yearly patterns rainfall of rainfall and also increases systematically over the entire Nkayi area although at different rates according to competition from other plant layers. As with shrubs, only goats will eat the tree biomass in height classes that are lower than 1.5 m in height. Once tree classes and their biomass have grown past this level, there is no further removal due to fires or domestic human use.

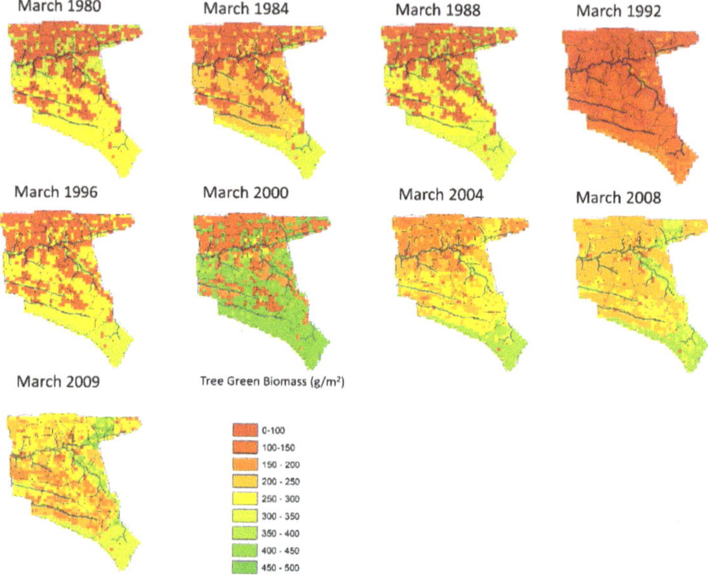

Figure 11. The spatial distribution of tree green biomass for selected years (March every 4th year of the historic scenario).

3.3. Historical Scenario: The Spatial Distribution of Shrub and Tree Cover

Figure 12 shows a comparison of tree and shrub cover maps for the years starting in 1980, 2000, and 2009. The increase in biomass is shown in Figure 10; Figure 11 above are also realized in systematic and pervasive bush encroachment by mostly woody tree species. Shrub cover does increase as well but in a patchier structure than woody plants. It is suspected that this dynamic is primarily due to the patchy initialization of shrub populations as well as goat browsing of the lower shrub layer and avoidance of the higher woody layers to browsing. Also note that the cover increase is not as linear as seen in the temporal biomass estimates seen in Figure 11. The spatial maps in Figure 12 show relatively slow cover increases in the first 10 to 20 years (1980–2000) and then faster increases over the next decade. This is illustrative of how local management choices can quickly alter the availability and quality of livestock forage.

3.4. Historical Scenario: Livestock Population Dynamics

SAVANNA simulations were executed with two sets of herbivore population demography options to explore the effect herbivore population on rangeland productivity. The first option was to set the livestock populations to constant levels' while the second option allowed livestock populations to rise and fall with effects of rangeland condition and water access. In both simulations, the fundamental simulation results in terms were similar that only the fluctuating populations are shown. Figure 13 shows the general livestock population dynamics simulated in most historic and climate change scenarios. Cattle, donkey, and goat populations increase slightly (within the yearly birth/death dynamics) until a dry period from the late 1980s until early 1990s causes a decline in all populations primarily due to grass available. Once higher rainfall amounts arrive in the middle to late 1990s, the three populations rise with cattle and donkey populations rising faster than goat populations. The rapid rise of donkey populations is most likely a function of fecundity/birth parameters and the lower initial population levels [34] Sheep populations do not change much for most of the simulation and then decline in the significant drought in 2002. This 2002 decline was also realized for goat populations and later for cattle populations. These later populations most likely show that more variable population dynamics are occurring as evidenced by the diminished grass forage resources and increased woody and shrub encroachment. These populations are generally within levels reported by [4] but do not have the detailed harvest dynamics described in these texts. The spatial distribution of livestock for selected years is shown in Figure 14.

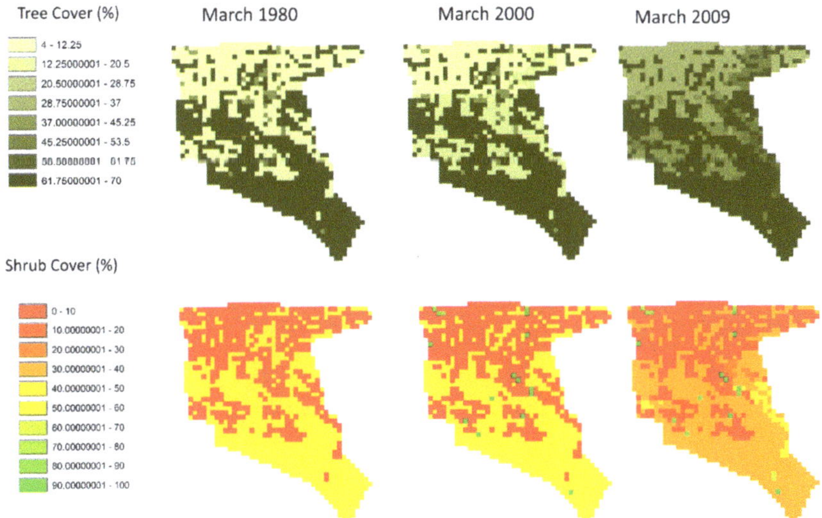

Figure 12. A comparison of the spatial distribution of tree and shrub cover for 1980, 2000 and 2009.

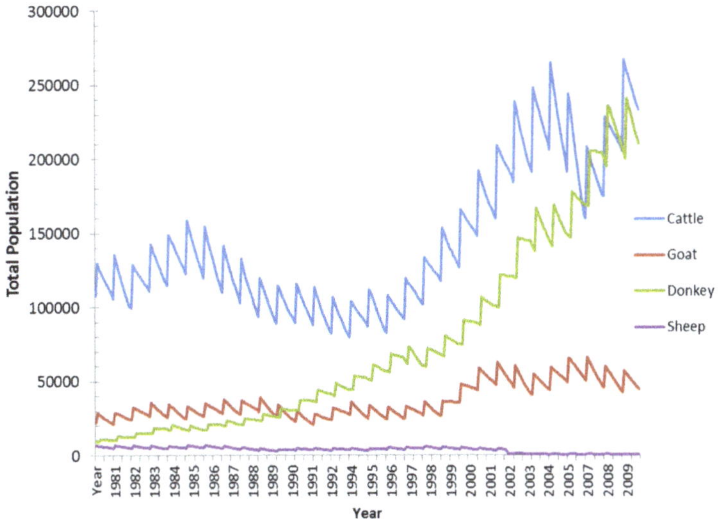

Figure 13. Simulated livestock populations for the Nkayi district in the historical scenario.

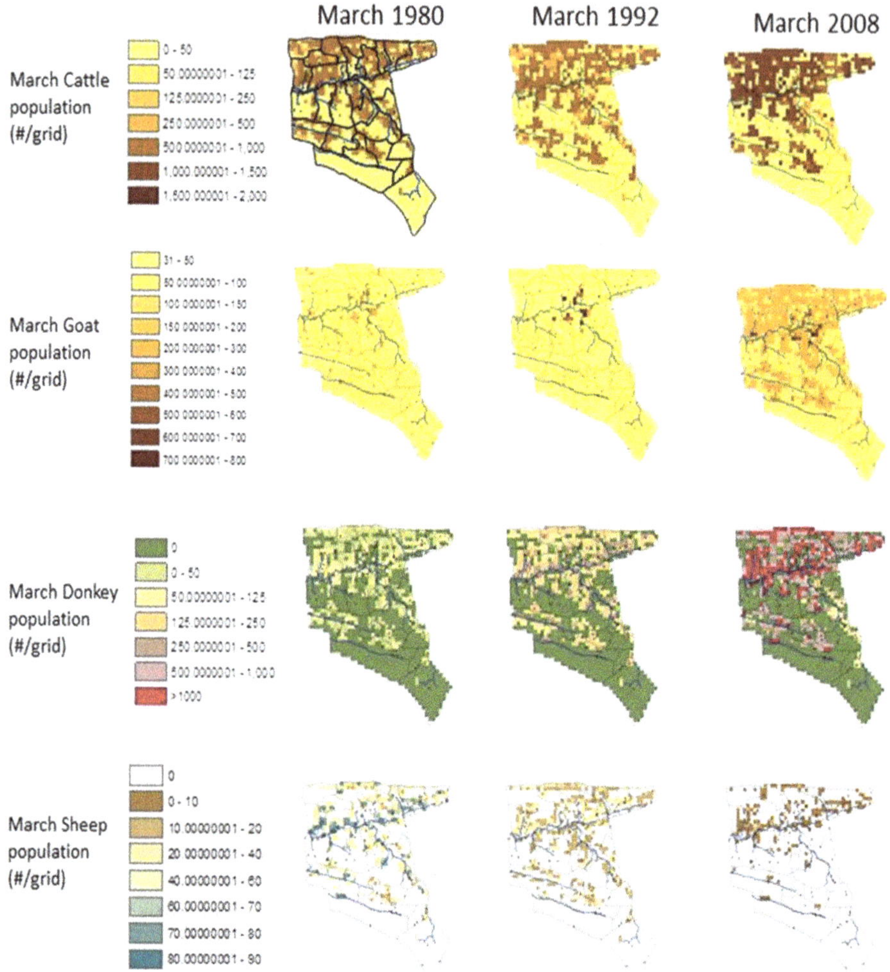

Figure 14. Spatial distribution of the livestock for selected years (March 1980, 1992 and 2008).

3.5. Climate Change Scenarios: Simulated Grass Biomass

Figure 15 shows simulated grass biomass (g/m^2) changing similarly in almost all climate scenarios, except for the peak and low years. These dynamics are mostly similar until about 2003 when the 650 ppm scenarios begin a systematic decrease. This decrease in grass production is probably due to increases in woody competition discussed below. The C4 photosynthetic pathway of the grass layer is generally less responsive to the increases in CO_2 levels. This decrease shows a systematic decrease in forage quantity and quality as the effects of livestock population increases and woody encroachment.

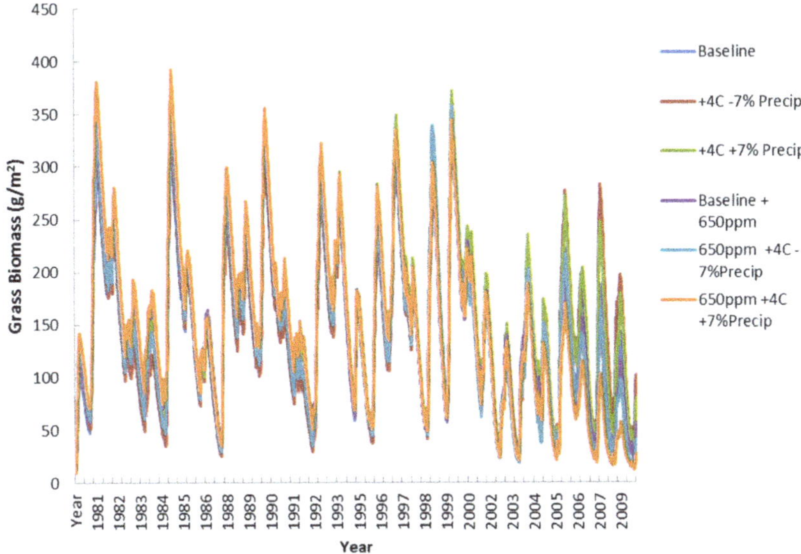

Figure 15. Simulated grass biomass (g/m^2) for all climate scenarios.

3.6. Climate Change Scenarios: Simulated Shrub Biomass

Figure 16 shows a similar systematic increase in shrub biomass as gained by its C_3 photosynthetic pathway. The increasing dynamics are similar to some changes in maximum or minimum levels. Shrub biomass levels are somewhat influenced by the CO_2 effects, but temperature and rainfall also feature prominently. More variation does occur in the second half of the simulation, but the overall increase is similar among all scenarios. These results have significant implications for the management of the woody shrub layer as grass forage availability may become limiting in later years with increased grazing pressure and changing climate conditions.

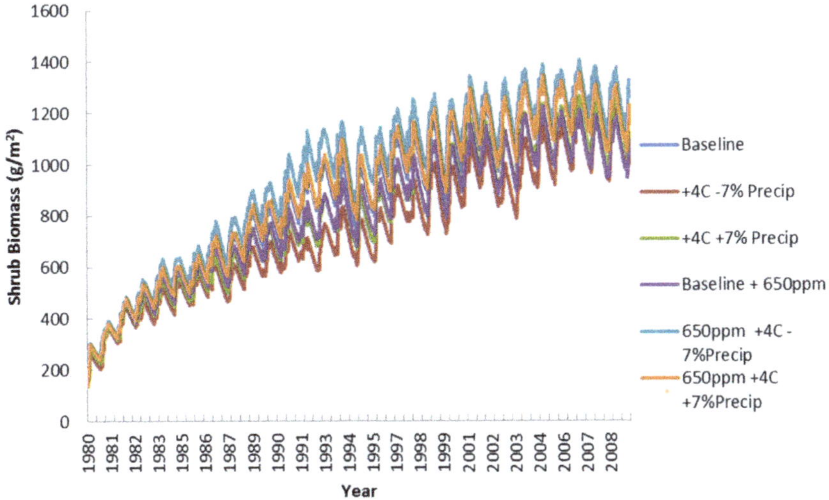

Figure 16. Simulated shrub biomass (g/m^2) for all climate scenarios.

3.7. Climate Change Scenarios: Simulated Tree Biomass

Figure 17 shows the largest differences among scenarios in comparison to the earlier grass and shrub figures. The tree biomass levels are significantly affected by the gains from elevated CO_2 and the assumption of a C_3 photosynthetic pathway. Interestingly, the three lower levels of woody biomass production are all the scenarios without the 650 ppm CO_2 level (Baseline, hot and wet, hot and dry) while the three higher biomass levels of woody biomass are with the 650 ppm of CO_2 (CO_2 + Baseline, CO_2 + hot and wet, CO_2 + hot and dry) These results show that the tree layer is more sensitive to elevated CO_2 than grass or shrub layers. This CO_2 sensitivity is even more important than temperature or rainfall conditions. In addition, the low palatability and height availability of the woody layer would mean lower herbivory and greater opportunity for woody encroachment. Given that fire is already suppressed in these savanna systems, the potential for expansion of the woody layer at the expense of the grass layer is significant.

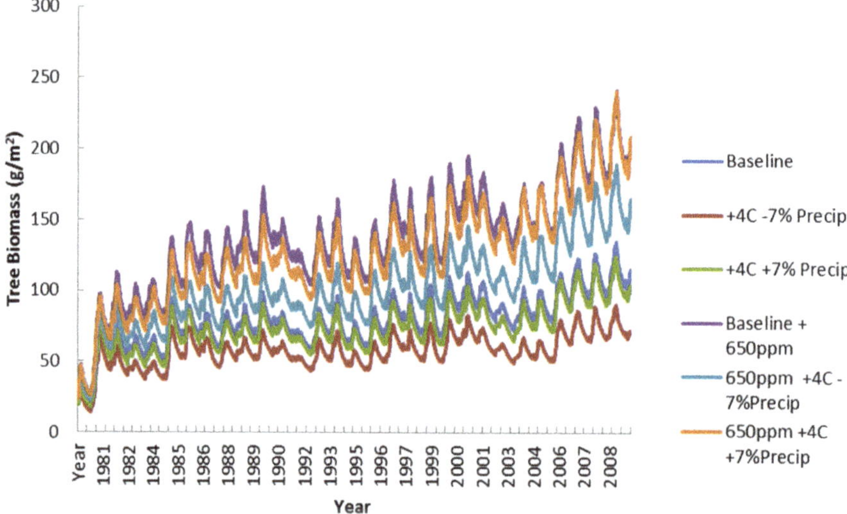

Figure 17. Simulated tree biomass (g/m^2) for all climate scenarios.

3.8. Climate Change Scenarios: Simulated Livestock Dynamics

In all scenario simulations of livestock, populations showed few variations in livestock populations that differed from base line. This trend continued for both fixed and for varying livestock populations. While initial conditions did change the nature of the final populations, additional sensitivity analysis simulations would be useful in determining whether livestock are truly as insensitive to climate conditions or whether additional parameterization is required.

4. Discussion

This section summarizes the different SAVANNA simulation results into four general discussion areas. SAVANNA has the advantage of both simulating temporal and spatial dynamics and both elements will be explored in these following sections. Savanna rangelands in Nkayi are characterized by a dynamics and potentially unstable mix of trees and grass growth forms driven by several interacting factors. Tree biomass can increase or decrease producing dense woodlands or open grasslands depending on the prevailing conditions. The trend has been that in many parts of the world there has been an increase in woody plant biomass and decline in open grasslands. The balance between the woody and herbaceous component influences livestock herbivory and an increase in the woody component is a major concern for livestock farmers. An increase in the woody component generally attributed to climate change and has the effect of reducing the grazing capacity.

4.1. Response of Herbaceous Biomass to Yearly Rainfall Totals

Herbaceous biomass responds mostly to yearly rainfall totals and tends to decrease with increase of woody species. This aspect is not surprising as most rangeland research has highlighted the importance of annual precipitation to seasonal herbaceous biomass production [10,11]. In 20 out of the 30-year simulation period, climate change scenarios differed from the baseline only in the magnitude of high and low biomass amounts while the overall trend followed multi-year climate trends. In the last decade of the scenarios, some divergence was seen, as increased woody plant encroachment and a larger herbivore population reduced herbaceous biomass. From a spatial perspective, herbaceous biomass is mostly distributed in the non-forest areas with significantly less production within forest areas with higher woody shrub and tree density as well as completion from these species. These results highlight that the grazing availability and overall quality may be significantly reduced by different combinations of low rainfall, woody plant competition and stocking rates. Additional simulations would aid in disentangling these various environmental and human-induced drivers.

4.2. Shrub Biomass and Cover Increase in All Historical and Climate Change Scenarios

This dynamic provides an important result and potential issue for future management consideration. Within the historical scenario, shrub biomass and spatial cover become more dominant over the grass and woody layer toward the mid-1980s to the 1990s, which was also reported in [10,44] The decrease of grass biomass and cover decreased continually as the shrubs increased in the latter part of the thirty-year simulation. This dynamic was quite stable under all climate and CO_2 scenario combinations with some small amount variation due to localized conditions. Spatially, the increase occurred in both the open lands as well as forested areas. Localized browsing of goats did have some effects on some grids but overall, more biomass and cover was created than consumed. Some of this increase may be mitigated by higher stocking rates of goats and other human bush reduction activities such as firewood collection.

4.3. Woody Biomass and Cover Increase Differently with Respect to Assumptions on Temperature, Precipitation and CO_2 Effects

This simulation result was probably the most surprising as woody growth diverged widely within the various climatic and CO_2 scenarios. While all woody growth and cover increased from initial levels, some scenarios (hot/dry/400 ppm, hot/wet/400 ppm, baseline/400 ppm) provided less increase than other scenarios (hot/wet/650 ppm, baseline/650 ppm, and hot/dry/650 ppm). In all these comparisons, the level of CO_2 was probably the most critical driver in the determination of woody growth. [33] cited that several authors proposed that woody plant expansion is associated with increases in the amount of carbon dioxide in the atmosphere over the past two decades. Other identified impacts of climate change on rangelands are increasing in bush encroachment and alterations in tree-grass interactions. Studies showed that increased carbon dioxide in the atmosphere may cause an increase in the woody layer [45,46]. They also suggest that the rising CO_2 levels may reduce the transpiration rates of grasses causing deeper percolation of water which in turn will support the establishment of tree seedlings and enhance growth of many trees. This will in turn increase soil water availability and hence competitive dominance and productivity of deep-rooted plants such as shrubs [33]. The same authors also showed that increased CO_2 in the atmosphere favors the post-fire regrowth of woody plants at the expense of grasslands. From the study carried out by Gordon [44], the trees were showed to predominate over the herbaceous biomass leading to bush encroachment. This was attributed it to earlier leaf emergence in trees and greater carbohydrate reserves. For the most part, the SAVANNA results agree with these research findings. Additional sensitivity analysis simulations with various parameter levels would help to explore these effects further.

4.4. Simulated Livestock Populations Are Not Particularly Sensitive to Climate Change Conditions

Another surprising result of the SAVANNA simulation is the relative insensitivity of livestock populations to the various climate scenarios. This effect was realized in both stable and set populations of livestock. In all scenarios, grazing herbivores were mostly distributed in the non-forest areas and near to water resources. This spatial limitation may have constrained livestock populations within the limited areas available. Forest areas were not significant resource areas for grazers as their herbaceous biomass was not particularly high, nor were they close enough to surface water to make the areas more favorable for occupation. Goats were the only livestock that were able to use the woodier areas for browsing. These results may be an artifact of model parameterization and further simulations would be warranted before assigning any additional confidence to model results. [10,44] showed that browse is an important feed resource for goats. They also mentioned that if feeding grass and browse become limiting, goats can expand their feed base to counter the effects of low feed quality and quantity. They again showed that cattle are less adaptable to seasonal changes of feed availability. The version of SAVANNA used does not simulate specific livestock management aspects such as moving cattle to specific wet and dry season grazing areas, nor does it simulate the limitation of communal areas due to crop production. Thus, livestock populations were distributed among all suitable areas (mostly in the non-forest areas) without regard to other human land uses. Thus, expanded SAVANNA and human activities simulations such as those by [26] would be useful to add more evidence to any conclusions toward livestock health under climate change conditions.

5. Conclusions and Recommendations

At the landscape scale, large herbivore–vegetation interactions can be quite complex involving many interacting factors [19]. Simulation modeling has proved a useful tool for disentangling some of this complexity [19,26]. From this study, trends in biomass production in Nkayi show that the herbaceous layer fluctuates around rainfall, while the woody layer increases at a slow but steady pace. On the other hand, the shrubs increased sharply over the years suggesting that there may be bush encroachment issues on the near horizon. Bush encroachment has a significant effect of reducing the grazing capacity and therefore stocking rates for cattle in Nkayi may need to be adjusted in the long run. Alternatively, farmers could increase off-takes to produce within the acceptable carrying capacity. Bush encroachment can be controlled by using species that are more of browsers than grazers, for example goats. Prescribed burning can also be used to control this potential encroachment. The study also concludes that the forest areas are not a critical feed source for cattle that are mainly grazers as they do not have enough herbaceous biomass (less than 50 g/m^2). Additionally, further SAVANNA simulations that include domestic firewood harvesting of shrub and tree layers in both forests and degraded lands would be valuable to include human adaptation mechanisms within this complex system.

It may be concluded from this initial study that climate change effects from the changing rainfall and temperature as predicted by the GCMs may not have strong negative effects on the rangelands near century. However, when considering that the rangelands are currently being systematically degraded from other anthropogenic factors and cannot support sustainable livestock production throughout the year, there is pressing need for a new strategy for agro-pastoral systems to enhance resilience and long-term sustainability.

If carbon dioxide levels continue to increase, the woody layer may respond more strongly than other vegetation layers. As such, there may be a need for farmers to invest in livestock species that are able to use the woody species. The biomass production trends during the year act as a guide for farmers to know when the critical feed shortages are, in terms of biomass production. This can help farmers to plan strategic supplementary feeding so as to avoid loss of body condition by the livestock. It is; however, critical to note that different livestock species have different feeding behaviors and responses to drought. This means that farmers need to plan their management and prioritize the most sensitive livestock species. For example, the 1992 drought had a huge impact on the sheep numbers. This means that those farmers that were using them as the main source of livelihoods were more adversely affected. Therefore, to hedge against such shocks in future farmers are encouraged to keep a mix if livestock species to potentially spread their risk.

This paper represents an initial simulation effort and not a final one as the results show a complex and coupled human/natural system. Local livelihoods are closely tied with climatic variability and livestock represent a significant and successful human mechanism for mitigating these variations. Whether the grazing system is fundamentally changing beneath resident's feet is a matter for continued discussion. Additional SAVANNA simulations would certainly help to disentangle the various human and environmental drivers to systematically address what human actions (de-stocking, changing from grazers to browsers, water availability) would add to overall resilience of these vulnerable agro-ecosystems.

Author Contributions: T.S.S. and G.A.K. both wrote the original manuscript conceptualization and draft. Additionally, they performed model calibration, parameterization and all simulations. P.M. and A.C. aided with model parameterization, and conceptualization as well as reviewing and editing the manuscript. J.v.N. provided review of the manuscript and academic supervision. All authors have read and agreed to the published version of the manuscript.

Funding: This research was funded by the Agriculture Model Inter-comparison and Improvement Project (AGMIP), at ICRISAT Bulawayo. Additionally, this work was funded in part by the United States Agency for International Development (USAID) Bureau for Food Security under Agreement # AID-OAA-L-15-00003 as part of Feed the Future Innovation Lab for Livestock Systems. Any opinions, findings, conclusions, or recommendations expressed here are those of the authors alone.

Acknowledgments: As authors we are grateful to AGMIP, USAID and ICRISAT for the financial support for undertaking this research. Our sincere gratitude to Dr Michael Coughenour for providing us with many of the livestock parameters as well as providing detailed SAVANNA insights. We are also grateful to the ICRISAT GIS lab for providing us with spatial information.

Conflicts of Interest: The authors declare no conflicts of interest.

References

1. Turner, M.D.; McPeak, J.G.; Ayantunde, A.A. The Role of Livestock Mobility in the Livelihood Strategies of Rural Peoples in Semi-Arid West Africa. *Hum. Ecol.* **2014**, *42*, 231–247. [CrossRef]
2. Seguin, B. *The Consequences of Global Warming for Agriculture and Food Production*; Rowlinson, P., Steele, M., Nefzaoui, A., Eds.; Cambridge University Press: Cambridge, UK, 2008; pp. 9–11.
3. Senda, T.S.; Peden, D.; Tui, S.H.-K.; Sisito, G.; Van Rooyen, A.F.; Sikosana, J.L.N. Gendered livelihood implications for improvements of livestock Water productivity In Zimbabwe. *Exp. Agric.* **2011**, *47*, 169–181. [CrossRef]
4. Goat Production and Marketing: Baseline Information for Semi-Arid Zimbabwe. Available online: http://oar.icrisat.org/397/1/CO_200704.pdf (accessed on 25 March 2020).
5. Pachauri, R.K.; Allen, M.R.; Barros, V.R.; Broome, J.; Cramer, W.; Christ, R.; Church, J.A.; Clarke, L.; Dahe, Q.; Dasgupa, P.; et al. *Climate Change 2014: Synthesis Report. Contribution of Working Groups I, II and III to the Fifth Assessment Report of the Intergovernmental Panel on Climate Change*; IPCC: Geneva, Switzerland, 2014; p. 151.
6. Gelats, F.L.; Fraser, E.D.; Morton, J.F.; Rivera-Ferre, M.G. What drives the vulnerability of pastoralists to global environmental change? A qualitative meta-analysis. *Glob. Environ. Chang.* **2016**, *39*, 258–274. [CrossRef]
7. Telesetsky, A. Organisation for Economic Co-operation and Development (OECD). *Int. Environ. Law* **2012**, *23*, 578–582. [CrossRef]
8. FAO and Platform for Agrobiodiversity Research. *Biodivers. Food Agric.* **2011**, *53*, 78.
9. Herrero, M.; Grace, D.; Njuki, J.; Johnson, N.; Enahoro, D.; Silvestri, S.; Rufino, M.C. The roles of livestock in developing countries. *Animal* **2012**, *7*, 3–18. [CrossRef]
10. Illius, A.W.; Gordon, I.J.; Derry, J.F.; Magadzire, Z.; Mukungurutse, E. Environmental variability and productivity of semi-arid grazing systems. In Proceedings of the Third Workshop on Livestock Production Programme Projects, Matobo, Zimbabwe, 28 September 2000; pp. 125–136.
11. Sullivan, S.; Rohde, R. On non-equilibrium in arid and semi-arid grazing systems. *J. Biogeogr.* **2002**, *29*, 1595–1618. [CrossRef]

12. Verburg, P.H.; Van De Steeg, J.; Veldkamp, A.; Willemen, L.; Veldkamp, T. From land cover change to land function dynamics: A major challenge to improve land characterization. *J. Environ. Manag.* **2009**, *90*, 1327–1335. [CrossRef]
13. Vetter, S. Rangelands at equilibrium and non-equilibrium: Recent developments in the debate. *J. Arid. Environ.* **2005**, *62*, 321–341. [CrossRef]
14. Boone, R.; Wang, G. Cattle dynamics in African grazing systems under variable climates. *J. Arid. Environ.* **2007**, *70*, 495–513. [CrossRef]
15. Puig, C.J.; Greiner, R.; Huchery, C.; Perkins, I.; Bowen, L.; Collier, N.; Garnett, S.T. Beyond cattle: Potential futures of the pastoral industry in the Northern Territory. *Rangel. J.* **2011**, *33*, 181–194. [CrossRef]
16. Masikati, P.; Tui, S.H.-K.; Descheemaeker, K.; Sisito, G.; Senda, T.S.; Crespo, O.; Nhamo, N. Integrated Assessment of Crop–Livestock Production Systems Beyond Biophysical Methods. In *Smart Technologies for Sustainable Smallholder Agriculture*; Elsevier BV: Amsterdam, The Netherlands, 2017; pp. 257–278.
17. Coughenour, M.B. Savanna-Landscape and Regional Ecosystem Model: User Manual. 1993. Available online: https://www.nrel.colostate.edu/assets/nrel_files/labs/coughenour-lab/pubs/Manual_1993/user_manual.pdf (accessed on 25 March 2020).
18. O'Connor, T.; Kiker, G.A. Collapse of the Mapungubwe Society: Vulnerability of Pastoralism to Increasing Aridity. *Clim. Chang.* **2004**, *66*, 49–66. [CrossRef]
19. Weisberg, P.J.; Coughenour, M.B.; Bugmann, H. *Large Herbivore Ecology, Ecosystem Dynamics and Conservation*; Cambridge University Press: Cambridge, UK, 2006; pp. 348–382.
20. Bunting, E.L.; Fullman, T.; Kiker, G.; Southworth, J. Utilization of the SAVANNA model to analyze future patterns of vegetation cover in Kruger National Park under changing climate. *Ecol. Model.* **2016**, *342*, 147–160. [CrossRef]
21. Hilbers, J.P.; Van Langevelde, F.; Prins, H.H.T.; Grant, C.C.; Peel, M.J.S.; Coughenour, M.B.; De Knegt, H.; Slotow, R.; Smit, I.P.J.; Kiker, G.A.; et al. Modeling elephant-mediated cascading effects of water point closure. *Ecol. Appl.* **2015**, *25*, 402–415. [CrossRef] [PubMed]
22. Fullman, T.J.; Bunting, E.L.; Kiker, G.A.; Southworth, J. Predicting shifts in large herbivore distributions under climate change and management using a spatially-explicit ecosystem model. *Ecol. Model.* **2017**, *352*, 1–18. [CrossRef]
23. Boone, R.; Galvin, K.A.; Coughenour, M.B.; Hudson, J.W.; Weisberg, P.J.; Vogel, C.H.; Ellis, J.E. Ecosystem Modeling Adds Value to a South African Climate Forecast. *Clim. Chang.* **2004**, *64*, 317–340. [CrossRef]
24. Thornton, P.; Fawcett, R.; Galvin, K.; Boone, R.; Hudson, J.; Vogel, C. Evaluating management options that use climate forecasts: Modelling livestock production systems in the semi-arid zone of South Africa. *Clim. Res.* **2004**, *26*, 33–42. [CrossRef]
25. Thornton, P.; BurnSilver, S.; Boone, R.; Galvin, K. Modelling the impacts of group ranch subdivision on agro-pastoral households in Kajiado, Kenya. *Agric. Syst.* **2006**, *87*, 331–356. [CrossRef]
26. Boone, R.; Galvin, K.A.; BurnSilver, S.B.; Thornton, P.; Ojima, D.S.; Jawson, J.R. Using Coupled Simulation Models to Link Pastoral Decision Making and Ecosystem Services. *Ecol. Soc.* **2011**, *16*. [CrossRef]
27. Boone, R.; Conant, R.T.; Sircely, J.; Thornton, P.; Herrero, M. Climate change impacts on selected global rangeland ecosystem services. *Glob. Chang. Boil.* **2017**, *24*, 1382–1393. [CrossRef]
28. Higgins, S.I.; Scheiter, S. Atmospheric CO_2 forces abrupt vegetation shifts locally, but not globally. *Nature.* **2012**, *488*, 209–212. [CrossRef] [PubMed]
29. Weber, U.; Jung, M.; Reichstein, M.; Beer, C.; Braakhekke, M.C.; Lehsten, V.; Ghent, D.; Kaduk, J.; Viovy, N.; Ciais, P.; et al. The interannual variability of Africa's ecosystem productivity: A multi-model analysis. *Biogeosciences* **2009**, *6*, 285–295. [CrossRef]
30. Kiker, G.A. Development and Comparison of Savanna Ecosystem Models to Explore the Concept of Carrying Capacity. Ph.D. Thesis, Cornell University, New York, NY, USA, August 1998.
31. Second Round Crop and Livestock Assessment Report: 2014/15 Season. 2015. Available online: https://documents.wfp.org/stellent/groups/public/documents/newsroom/wfp275316.pdf (accessed on 25 March 2020).
32. Stuart-Hill, G.; Scholes, R.J.; Walker, B.H. An African Savanna: Synthesis of the Nylsvley Study. *J. Appl. Ecol.* **1994**, *31*, 791. [CrossRef]
33. Bond, W.J.; Midgley, G.F. A proposed CO2-controlled mechanism of woody plant invasion in grasslands and savannas. *Glob. Chang. Boil.* **2000**, *6*, 865–869. [CrossRef]

34. Michael, B.C. Natural Resource Ecology Laboratory, Colorado State University. Unpublished work. **2014**.
35. Warf, B.; Poore, B.S. United States Geological Survey (USGS) (2020) Earth Explorer. Available online: https://earthexplorer.usgs.gov/ (accessed on 25 March 2020).
36. Chirima, A.; Vanrooyen, A.; Homann, S.; Mundy, P.; Ncube, N.; Sibanda, N. Land degradation: Evaluating land cover change of mixed crop livestock systems in the semi-arid Zimbabwe from 1990 to 2009. Midlands State Univ. *J. Sci. Agric. Technol.* **2012**, *3*.
37. Whitlow, R. Potential versus actual erosion in Zimbabwe. *Appl. Geogr.* **1988**, *8*, 87–100. [CrossRef]
38. Pike, A.; Schulze, R.E. *AUTOSOIL Version 3: A Soils Decision Support System for South African Soils*; Department of Agricultural Engineering, University of Natal: Pietermaritzburg, South Africa, 1995.
39. Soils: Agrohydrological Information Needs, Information Sources and Decision Support. Available online: http://sarva2.dirisa.org/ (accessed on 25 March 2020).
40. Senda, T.S. University of Nairobi, LARMAT, Kenya. Unpublished work. **2020**.
41. Antle, J.; Valdivia, R. Modelling the supply of ecosystem services from agriculture: A minimum-data approach. *Aust. J. Agric. Resour. Econ.* **2006**, *50*, 1–15. [CrossRef]
42. Gama, A.C.; Mapemba, L.D.; Masikati, P.; Tui, S.H.K.; Crespo, O.; Bandason, E. *Modeling Potential Impacts of Future Climate Change in Mzimba District, Malawi, 2040–2070: An Integrated Biophysical and Economic Modeling Approach*; International Food Policy Research Institute: Washington, DC, USA, 2014; Volume 8.
43. Coughenour, M. The Ellis paradigm—humans, herbivores and rangeland systems. *Afr. J. Range Forage Sci.* **2004**, *21*, 191–200. [CrossRef]
44. Gordon, I.J. Browsing and grazing ruminants: Are they different beasts? *For. Ecol. Manag.* **2003**, *181*, 13–21. [CrossRef]
45. Forests, Rangelands and Climate Change in Southern Africa. Available online: https://www.uncclearn.org/sites/default/files/inventory/fao190.pdf (accessed on 25 March 2020).
46. Mokgosi, R.O. Effects of bush encroachment control in a communal managed area in the Taung region, North West Province, South Africa. Ph.D. Thesis, North-West University, Potchefstroom, South Africa, May 2018.

Article

An Evaluation of Vegetation Health in and around Southern African National Parks during the 21st Century (2000–2016)

Hannah Herrero [1,*], Jane Southworth [2], Carly Muir [2], Reza Khatami [2], Erin Bunting [3] and Brian Child [2]

[1] Department of Geography, University of Tennessee, 1000 Philip Fulmer Way, Room 315, Knoxville, TN 37996-0925, USA

[2] Department of Geography, University of Florida, 3141 Turlington Hall, Gainesville, FL 32611, USA; jsouthwo@ufl.edu (J.S.); carlysmuir@ufl.edu (C.M.); seyedghkhatami@ufl.edu (R.K.); bchild@ufl.edu (B.C.)

[3] Department of Geography, Environment, and Spatial Sciences; Michigan State University, 673 Auditorium Rd., East Lansing, MI 48825, USA; ebunting@msu.edu

* Correspondence: hherrero@utk.edu; Tel.: +1-865-974-6043

Received: 28 February 2020; Accepted: 26 March 2020; Published: 30 March 2020

Abstract: Roughly 65% of the African continent is classified as savanna. Such regions are of critical importance given their high levels of biological productivity, role in the carbon cycle, structural differences, and support of large human populations. Across southern Africa there are 79 national parks within savanna landscapes. Understanding trends and factors of vegetation health in these parks is critical for proper management and sustainability. This research strives to understand factors and trends in vegetation health from 2000 to 2016 in and around the 79 national parks across southern Africa. A backward stepwise regression was used to understand the factors (e.g., precipitation, population density, and presence of transfrontier conservation areas) affecting the normalized difference vegetation index (NDVI) during the 21st century. There was a statistically significant positive ($p < 0.05$) relationship between mean annual precipitation and NDVI, and a significant negative relationship between population density and NDVI. To monitor vegetation trends in and around the parks, directional persistence, a seasonal NDVI time series-based trend analysis, was used. Directional persistence is the net accumulation of directional change in NDVI over time in a given period relative to a fixed benchmarked period. Parks and buffer zones across size classes were compared to examine differences in vegetation health. There was an overwhelmingly positive trend throughout. Additionally, national parks, overall, had higher amounts of positive persistence and lower amounts of negative persistence than the surrounding buffer zones. Having higher positive persistence inside of parks indicates that they are functioning favorably relative to the buffer zones in terms of vegetation resilience. This is an important finding for park managers and conservation overall in Southern Africa.

Keywords: remote sensing; vegetation dynamics; savanna landscapes; national parks; conservation

---

1. Introduction

Globally, drylands cover around 50% of the earth's terrestrial surface [1]. Of these drylands, savannas cover extensive portions of Africa, South America, and Australia. In southern Africa, over 55% of the landscape is classified as savanna [1,2]. Savannas are a type of dryland, and therefore are water limited systems. They are traditionally defined as grasslands with scattered trees, but practically are defined as a heterogeneous landscape composed of a mix of grasses, shrubs, and trees [1,3,4]. Savannas play a key role in the global carbon cycle and global net primary production [5–10]. Due to their importance and high levels of utilization, it is critical to understand savanna landscape patterns over time for management of these systems, especially into the future under climate change. The major drivers of savanna systems are precipitation, fire, herbivory, and humans [11–14]. Parks and protected areas are used extensively across landscapes to conserve and protect the natural vegetation and wildlife populations. Countries across southern Africa employ varied management approaches within such protected areas, as regards fire, herbivory, and human usage. These management practices range from no fire management to regular prescribed burns, no animal population control to culling of herds to maintain ideal population sizes, and allowing extractive use of resources to not permitting any human use of natural resources [15]. As such, understanding the success of such different and varied protection regimes is key.

Using remote sensing to monitor vegetation in savannas, while important, is notoriously difficult, as these landscapes are transitional with variable vegetation cover. Because a savanna can be any structure of vegetation in the gradient of grassland, shrubland, or woodland, certain continuous techniques have been developed to best study these systems over time. A significant study in the literature concerning drylands has developed a novel approach, utilizing continuous time series analyses, to show change in these systems over time [16]. Time series analysis of the normalized difference vegetation index (NDVI) is often used as a proxy for vegetation health and abundance [16–22]. Using such time series approaches, we can draw information about vegetation properties and ecological responses to change at a regional scale [23]. One result that has been produced in the literature is that of "greening". Using satellite remote sensing, global greening has been observed during the 21st century by the National Aeronautics and Space Administration (NASA) and the National Oceanic and Atmospheric Administration (NOAA) [24]. Several studies have concluded that CO_2 fertilization is the largest contributing factor. Other factors include nitrogen levels, land use, temperature, and climate [25–27].

Precipitation is known as the predominant factor in shaping landscape pattern and productivity across southern Africa [9]. In areas with up to mean annual precipitation of 750 mm, grassland dominated systems are expected [28]. More mixed systems are expected for areas with precipitation between 750 and 950 mm. For areas with precipitation above 950 mm, trees become the more common vegetation type [9,14,28,29]. Further, Campo-Bescos et al. (2013) and others found that the boundary between a mixed system and tree-dominated system is influenced by temperature [28]. Specifically, temperature and resulting decreases in soil moisture can adversely impact vegetation health and productivity in savanna systems [30]. Access to deeper ground water can also play a role in a tree's ability to exist on the landscape [31]. Population pressure and human utilization of the landscape are increasing, putting greater pressure on protected areas through land conversion (positive impact) and extractive use of resources (negative impact) [2,32,33]. Elevation may also play a role in the health of vegetation, as elevation can constrain the growth of plants due to topographic climate effects [34].

In this study, we focus on the 79 national parks in southern Africa, all of which are some variation of savannas, varying from sparse grassland to deciduous woodland cover types. These parks are of critical importance, both socioeconomically and ecologically, as they support large human populations though tourism revenue and natural resource extraction, and support high levels of biodiversity [35,36]. Parks are important conservation features on the landscape because they tend to focus on maintaining biodiversity in vegetation and wildlife while minimizing human impacts. Theories in landscape ecology suggest that larger parks tend to be more successful in achieving their conservation goals [37], but the land matrix surrounding the park is also important. A way to better understand this matrix is to develop a comparison between the park and the surrounding matrix, or buffer zone. Buffer zones are important in their role of alleviating human impact on national parks [38]. More recently, multiple country, cross-border parks have also been developed, which are known as transfrontier conservation areas. The creation of transfrontier conservation areas is based on the landscape ecology principles of connectivity—the more connected, the better for conserving biodiversity (positive impact; [39]). It is now an important conservation question as to whether such transfrontier conservation areas are actually working and, as such, to address the landscape ecology theory that the size and connectivity of a protected area impacts the overall quality [37].

Currently, conservation development via parks and protected areas is a global phenomenon. However, such landscapes have rarely been studied on a larger scale explicitly to determine the factors of vegetation greenness or park success. This research examines factors of spatial heterogeneity in NDVI, specifically addressing the role of parks versus buffer zones or the surrounding land matrix. Such potential factors of savanna vegetation within these protected areas and their surrounding buffers include precipitation, human population density, transfrontier conservation areas, elevation, park size and area, land cover, and country. The research objectives of this study are as follows: (1) to understand the factors of vegetation health in national parks and their associated buffer zones in southern Africa; and (2) to determine how savanna vegetation persistence within these national parks and buffer zones of southern Africa varied from 2000 to 2016.

2. Materials and Methods

2.1. Study Area

For this study, 79 national parks across eight countries (South Africa, Zambia, Zimbabwe, Botswana, Namibia, Mozambique, Malawi, and Angola) in southern Africa were analyzed (Figure 1). Of the various protected area designations (e.g., national parks, game management areas, private game reserves), national parks were selected for their consistency in definition across countries. These parks were taken from the World Database on Protected Areas [40]. Given the spatial scale of the remote sensing techniques, only parks greater than 312,500 m^2 were studied because five or more pixels at the 250 m spatial resolution were needed to be able to establish trends. For this reason, the national parks of Swaziland and Lesotho were not included in this study. Given the large variation in area across these 79 parks, using a natural breaks statistical technique, parks were broken into three main size classes: small, medium, and large (Figure 2). The small size class had a sample size of 41 and a mean area of 302 km^2. The medium size class had a sample size of 30 and a mean area of 4,771 km^2. The large size class had a sample size of 8 and a mean area of 23,284 km^2 (Figure 2). Buffer zones with sizes of 10 km, 20 km, and 30 km were created for the small, medium, and large parks, respectively. Buffers were created to be able to compare vegetation persistence inside and outside of the park. Buffer sizes chosen for each class are based on literature suggesting that buffers be approximately three times the size of the park [38]. Across this study region, climate varies greatly and was accounted for by comparing mean annual precipitation from the Global Historical Climate Network [41]. Mean annual precipitation of parks in this study varied from 52 to 1466 mm. Land cover variations were defined as savannas with varying amounts of grass/wood biomass. These differences were taken into account by using the Global Land Project Facility data on global land cover [42] (Table A1). Of these 79 national parks, 31 are considered to be transfrontier conservation areas, which can be defined as continuous conservation areas divided by geopolitical boundaries (namely country boundaries) that are cooperating together, but may still function under different management regimes [43].

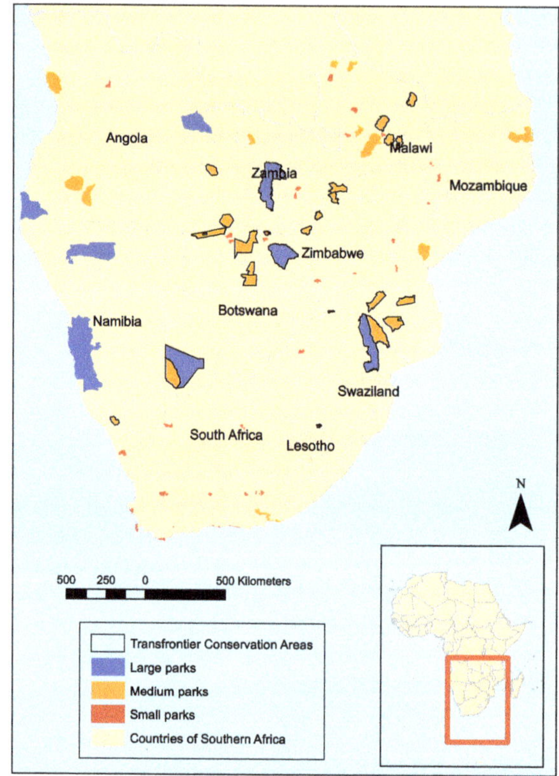

Figure 1. Southern Africa with national parks across the landscape and transfrontier parks highlighted.

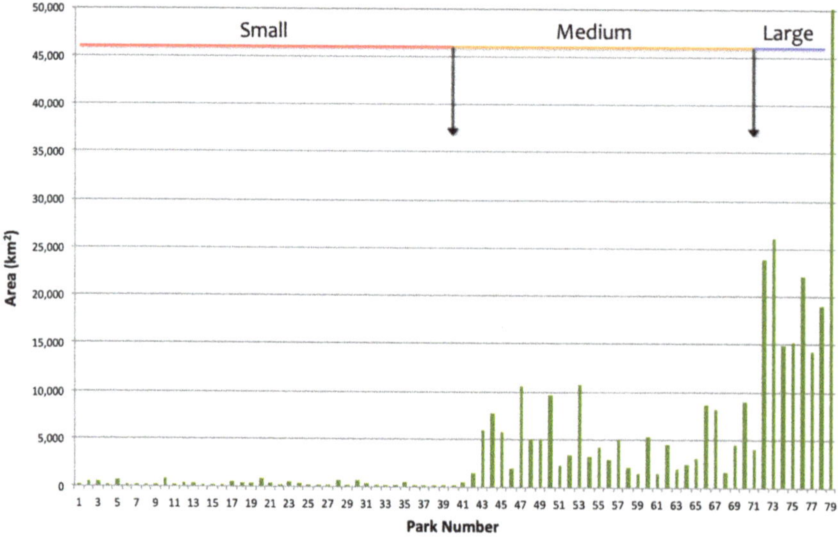

Figure 2. National parks used in the study by area, showing the breakdown of the three size area classes: small, medium, and large.

2.2. Understanding Factors of Vegetation Cover: Stepwise Backward Regression

This section addresses factors used to measure vegetation change and related factors.

2.2.1. Model Selection

A backward stepwise regression was used for this analysis because this method is capable of selecting variables with the most explanatory power from a number of input variables [44]. All assumptions of linear models were tested before employing the model: a linear relationship between dependent and independent variables, statistical independence of the errors, homoscedasticity of the errors, and confirmation that the error distribution was normal. This model has a relatively higher level of robustness for small sample sizes [44] compared to a geographically weighted regression. However, to account for the park's spatial distribution on the landscape, we used variables that address differences across space, such as land cover, elevation, and mean annual precipitation. The model starts with all possible variables, a "full" model, and then runs several times to eliminate variables that are deemed not significant by the model. This is done by starting with all variables and then testing each variable by deleting it from the model. Variables are continuously deleted until the loss gives the least significant deterioration with the highest model fit. The final model of significant variables is referred to as a "reduced" model. The Akaike Information Criterion (AIC) is a technique that is based on in-sample fit to then estimate how well a model does at predicting future values [45]. Between models, the one with the lowest AIC is determined to be the model of best fit. Therefore, between the range of full and reduced models, the lowest AIC was used to select the best model here.

2.2.2. Model Variables

In order to understand the factors of vegetation health in and around national parks of southern Africa, the normalized difference vegetation index (NDVI) was used as a measure of vegetation health and abundance. NDVI is calculated as: $\frac{NIR-R}{NIR+R}$, where R is the red band and NIR is the near infrared band [46]. This index works on the principle that NIR is highly reflective in healthy vegetation and red is highly absorptive in healthy vegetation. NDVI values range from −1 and +1, with any value above 0 indicating vegetation and above 0.9 saturating out in dense healthy vegetation [46]. The NDVI product from the Moderate Resolution Imaging Spectroradiometer (MODIS) MOD13Q1.006 Terra Vegetation Index (250 m spatial resolution) was used in this research. This NDVI product was obtained from Google Earth Engine (GEE) and had been atmospherically corrected on a per-pixel basis and was also pre-masked for water, clouds, heavy aerosols, and cloud shadows. The data has also been through the quality assessment check with the provided pixel reliability band. These products are 16 day MODIS observation composites where the pixel value is equal to the NDVI value of the observation closest to nadir view. This is determined between the top two NDVI observations during a 16 day period. The monthly product used spanned from March 2000-November 2016. This data was aggregated into seasons: December, January, February (DJF)—boreal winter/southern Africa wet season; March, April, May (MAM)—boreal spring/southern Africa wet season; June, July, August (JJA)—boreal summer/southern Africa dry season; and September, October, November (SON)—boreal autumn/southern Africa dry season. This study focuses only on the wet seasons (DJF and MAM), given that during the dry season, NDVI values in this region are very low. Over the 21st century, these seasons were broken into finer temporal periods of 2000–2004, 2005–2008, 2009–2012, and 2013–2016 to determine if different factors were important at different times across the overall time period of 2000–2016. The seasonal mean NDVI during each time period was used as the dependent variable in the following regressions.

The independent variables in these regressions were: country (X1), land cover (X2), area (X3), mean annual precipitation (X4), human population (X5), elevation (X6), and transfrontier (X7). All of the variables except for country (coded) and transfrontier (yes or no) were at the 1 km spatial resolution, and were extracted at the park/buffer scale. The country variable is designated in the model to account for management type, depending on the country name and boundaries, which were obtained from the Thematic Mapping website [47]. The land cover variable is designated in the model depending on the cover type [42]. Within a park or buffer, the land cover type at a pixel level occurring most often was used as a representation of the respective area. Generally, land covers across these national parks are varying types of savannas [1]. This land cover data was obtained from the Global Land Cover Facility at the University of Maryland [42]. The area variable represented the size of each park and buffer zone. Precipitation, gathered from the Global Historical Climatology Network, represents longer-term climate, and was calculated by creating the mean annual precipitation (MAP) across the parks and buffers over 30 years [41]. The population density variable is measured in people per km^2 and was obtained from the WorldPop 2015 dataset [48]. Elevation, which has a complex relationship with vegetation (at extreme highs there is little vegetation, but at mid-level vegetation trees may be the most productive vegetation types and grasses may be less productive, and shrubs may decline consistently with increasing vegetation) is potentially an important explanatory variable [49]. The average elevation was calculated (m) for each park and buffer with data from the Shuttle Radar Topography Mission [50]. The transfrontier variable is binary and the list of transfrontier parks were taken from the Peace Parks Foundation, which is key in setting up and managing these areas [43].

2.2.3. Vegetation Directional Persistence

In addition to the analysis of NDVI absolute values, we investigated the per-pixel time series NDVI values using a measure called vegetation directional persistence. Directional persistence utilizes a novel time series approach based on vegetation indices, such as NDVI, to determine the overall health and persistence of vegetation cover over time, in this study, from 2000 to 2016 [23]. Directional persistence (D) is a cumulative direction of change over the time period of interest when compared to a benchmark observation of NDVI [16,23]. Using benchmark NDVI values and comparison NDVI values in years after the benchmark has been shown to be an indicator of vegetation trends over time and space [16,23]. A random walk statistic is applied to each pixel highlighting statistically significant increase or decrease in vegetation persistence, or health of vegetation at a pixel level [23]. This statistic is based on the accumulation of consecutive outcomes during a Bernoulli process. During this process the probability of a success (step = +1) and a failure (step = −1) are independent of the position of the random walk at that time so the probabilities of a success or failure are not dependent on previous successes or failures [23]. The MODIS NDVI data, described previously, was used for the directional persistence analysis by season (DJF, MAM, JJA, SON). The benchmark seasonal mean NDVI values were obtained for the start of the time series of MODIS data, from 2000 to 2004. This accounts for any anomalous years and corresponds with the timeline from the regression analysis in the first part of this study. The latter part of the NDVI time series was the comparison period from 2005 to 2016 (n = 12). For each season, the mean NDVI pixel value of each of the comparison years was compared with the mean NDVI pixel value from the benchmark [23], for every pixel in the image. If the value in a given comparison year was higher than the benchmark, the pixel was assigned a value of +1, and if the value in a given comparison year was lower than the benchmark, the pixel was assigned a −1. These values for the 12 years were then accumulated to produce a net value of D for each pixel. Because this method uses a random walk statistic, a statistical significance can be attached to each pixel [23]. Given this is a two-tailed test, for the significance level of 0.05, pixels with values of +/−8 were considered to have a significant trend in either a positive or negative direction. For the purposes of this study, directional persistence values were mapped across the entire region of southern Africa, but overall percentages of positive and negative significant values were extracted inside of parks, as well as inside of buffer zones, so a direct comparison between these two landscapes could be made.

3. Results

3.1. Stepwise Backward Regression

Results of the stepwise backward regression are presented below in Table 1a,b, and Table 2a,b. Each of the models, regardless of season, time period, and park or buffer, had over 60% of the variation explained by the independent variables. This indicates that the models are successful in explaining the majority of spatial variations in NDVI values. However, parks in each season and period had higher R^2 values than buffers. There was also a greater range in park R^2 values relative to buffers. For parks, between 65% and 69% of the NDVI variation was explained by these models across seasons and time periods. For buffers, between 61% and 62% of the NDVI variation was explained by these models across seasons and time periods. The model varied slightly in the number of variables in the best model, but across all models, country (X1) was never used as an important variable. The intercepts in these models were very low (hovering around zero) due to the small range in which NDVI is measured. The Akaike Information Criterion (AIC) was included to demonstrate exactly how much these models differed, but within each of the season, time periods, and parks or buffers, the lowest AIC was used to select the best model. In comparison across parks and buffers, the parks had lower AIC values, also supporting that the park models performed better overall.

Table 1. Stepwise backward regression results including the best model, intercept, and Akaike Criterion for parks and buffers at each of the time periods 2000–2004, 2005–2008, 2009–2012, 2013–2016 in (**a**) December, January and (**b**) February, March, April, and May. The independent variables in these regressions were: country (X1), land cover (X2), area (X3), mean annual precipitation (X4), human population (X5), elevation (X6), and transfrontier (X7 and n = 31 of the 79 parks).

(**a**)

Area/Time	R^2	Best Model	Intercept	Akaike's Criterion
Parks DJF 2000–2004	0.68	y~X2 + X4 + X5 + X7	2.346×10^{-1}	−331.36
Buffers DJF 2000–2004	0.62	y~X2 + X4 + X5 + X6 + X7	-4.158×10^{-2}	−276.89
Parks DJF 2005–2008	0.67	y~X3 + X4 + X5 + X7	1.813×10^{-1}	−326.71
Buffers DJF 2005–2008	0.62	y~X2 + X4 + X5 + X6 + X7	-3.596×10^{-2}	−275.83
Parks DJF 2009–2012	0.66	y~X4 + X5 + X7	1.656×10^{-1}	−325.82
Buffers DJF 2009–2012	0.61	y~X2 + X4 + X5 + X6 + X7	-2.330×10^{-2}	−272.69
Parks DJF 2013–2016	0.65	y~X2 + X3 + X4 + X5 + X7	2.502×10^{-1}	−317.62
Buffers DJF 2013–2016	0.62	y~X2 + X4 + X5 + X6 + X7	-3.360×10^{-2}	−277.28

(**b**)

Area/Time	R^2	Best Model	Intercept	Akaike's Criterion
Parks MAM 2000–2004	0.69	y~X2 + X4 + X5 + X6 + X7	3.383×10^{-1}	−350.74
Buffers MAM 2000–2004	0.61	y~X2 + X4 + X5 + X7	3.079×10^{-2}	−293.31
Parks MAM 2005–2008	0.69	y~X2 + X4 + X5 + X6 + X7	3.004×10^{-1}	−352.89
Buffers MAM 2005–2008	0.62	y~X2 + X4 + X5 + X7	3.756×10^{-2}	−297.45
Parks MAM 2009–2012	0.67	y~X2 + X4 + X5 + X7	3.108×10^{-1}	−348.22
Buffers MAM 2009–2012	0.62	y~X2 + X4 + X5 + X7	3.090×10^{-2}	−293.39
Parks MAM 2013–2016	0.69	y~X2 + X4 + X5 + X6 + X7	3.363×10^{-1}	−343.49
Buffers MAM 2013–2016	0.62	y~X2 + X4 + X5 + X7	3.547×10^{-2}	−292.97

Table 2. Stepwise backward regression results for parks and buffers in (**a**) December, January, February and (**b**) March, April, and May at each of the time periods 2000–2004, 2005–2008, 2009–2012, 2013–2016 including each model independent variable: country (X1), land cover (X2), area (X3), mean annual precipitation (X4), human population (X5), elevation (X6), and transfrontier (X7 and n = 31 of the 79 parks), positive or negative relationship between the dependent and independent variable (if applicable), and the level of significance of the relationship between the dependent and independent variable. Variable significance is designated by asterisks. More asterisks indicate higher significance: 0 = **** 0.001 = *** 0.01 = ** 0.05 = * 1 = none.

(a)

Area/Time	Country	Land Cover	Area (km^2)	MAP (mm)	Mean Population	Mean Elevation	Transfrontier (Y/N)
Parks DJF 2000–2004				+ ****	− *		
Buffers DJF 2000–2004		*		+ ****	− ****	+ *	+ ***
Parks DJF 2005–2008			− *	+ ****	− *		
Buffers DJF 2005–2008		*		+ ****	− ****		+ ***
Parks DJF 2009–2012				+ ****			+ ***
Buffers DJF 2009–2012		*		+ ****	− ****		+ ***
Parks DJF 2013–2016				+ ****			+ ***
Buffers DJF 2013–2016				+ ****	− ****	+ *	+ ***
Variable Significance: 0 = ** 0.001 = *** 0.01 = ** 0.05 = * 1 = none**							

(b)

Area/Time	Country	Land Cover	Area (km^2)	MAP (mm)	Mean Population	Mean Elevation	Transfrontier (Y/N)
Parks MAM 2000–2004		**		+ ****	− **	− **	+ **
Buffers MAM 2000–2004		**		+ ****	− ****		+ **
Parks MAM 2005–2008		*		+ ****			+ **
Buffers MAM 2005–2008		**		+ ****	− ****		+ **
Parks MAM 2009–2012		**		+ ****			+ **
Buffers MAM 2009–2012		***		+ ****	− ****		+ **
Parks MAM 2013–2016		**		+ ****	− *	− *	+ **
Buffers MAM 2013–2016		**		+ ****	− ****		+ ***
Variable Significance: 0 = ** 0.001 = *** 0.01 = ** 0.05 = * 1 = none**							

A variable results table of the stepwise backward regression are below in Table 1a,b, and Table 2a,b. This shows each of the variables, when they were significant, how they were significant (positive or negative, if applicable), and what statistical significance level ($p < 0.001$ = ****, 0.001 = ***, 0.01 = **, 0.05 = *, or no * for no significance).

In every model, the most important independent variable was MAP. MAP had a positive relationship with NDVI, meaning that higher precipitation resulted in higher NDVI values. MAP consistently had the highest level of significance (Table 1a,b, and Table 2a,b, <0.001) across each model. This is expected considering that savannas are water-limited systems.

The second most important independent variable was mean population density. Mean population density had a negative relationship with NDVI. This means where there were higher human populations, there were lower NDVI values, likely due to human use of the landscape and potential subsequent degradation. Mean population density was significant across most of the models, but tended to be most significant in the buffer zones (Table 1a,b, and Table 2a,b, <0.001). This is due to higher densities of people in the buffer zones than in the parks. In parks, mean population density was significant at a relatively lower significance level and only in DJF 2000–2004 (0.05) and 2005–2008 (0.05), and in MAM 2000–2004 (0.01) and 2013–2016 (0.05). The relationship between population density and vegetation health is weaker in parks than in buffer zones. Over time, the level of significance has decreased, as parks have become better protected.

Another important variable was whether or not a park was part of a transfrontier conservation area. Transfrontier areas had a positive relationship with NDVI values, so where there was a transfrontier conservation area, NDVI values were higher. This might be expected because areas of cooperation may lead to better management, with healthier vegetation. This variable had a high level of significance (Table 1a,b, and Table 2a,b, 0.001) across parks and buffers in DJF, except for parks in the earliest periods (2000–2004 and 2005–2008) when these transfrontier areas were not yet strongly established. Across parks and buffers in MAM, transfrontier areas had a mid level of significance (Table 1a,b, and Table 2a,b, 0.01), except for buffers in 2013–2016, which was a higher level of significance (Table 1a,b, and Table 2a,b, 0.001). This result is likely showing a strong significance due to the firm establishment of these transfrontier areas in this period, which helps to mitigate some human impact. This may be combined with the high level of biomass at the end of the rains then, when NDVI values are at a maximum.

Elevation had a complicated relationship with NDVI. Mean elevation had a positive relationship with NDVI only in the buffer zones of DJF in 2000–2004 and 2013–2016 (0.05 level of significance), so with higher elevation there were higher NDVI values. This may be caused by humans not utilizing the landscape at higher elevations. Mean elevation had a negative relationship with NDVI only in parks of MAM in 2000–2004 (0.01) and 2013–2016 (0.05), so with higher elevation there were lower NDVI values. However, this may be due to a lack of vegetation at higher elevation in some parks. Overall, this was a much more mixed and weaker result.

Area was not as significant as was hypothesized prior to the analysis. It was only significant in parks at the 0.05 level in DJF 2005–2008 (Table 1a,b, and Table 2a,b). Area had a negative relationship to NDVI, meaning for this one model, the larger the park the lower the NDVI values, which was unexpected.

Land cover was significant in some cases. In DJF in buffer zones from 2000–2004, 2005–2008, and 2009–2012 land cover was significant at the 0.05 level. In MAM, land cover was significant (varying between 0.05, 0.01, and 0.001) across parks and buffers in all time periods. This suggests that land cover plays a more important role in NDVI values later in the wet season when the NDVI values themselves are much higher.

3.2. Vegetation Directional Persistence

Directional persistence varied spatially across southern Africa and across seasons (Figure 3a,b, all values of persistence; and Figure 3c,d, significant values only). In DJF there was a strong positive trend in the western portion of southern Africa, most strongly in Namibia and Angola. The area near the borders of South Africa, Zimbabwe, Botswana, and Zambia also had a very strong positive trend in persistence, which is one of the largest transfrontier conservation areas in southern Africa. The most northern region of the study area had neutral persistence. In eastern Mozambique and western Angola there was a weak negative trend in persistence.

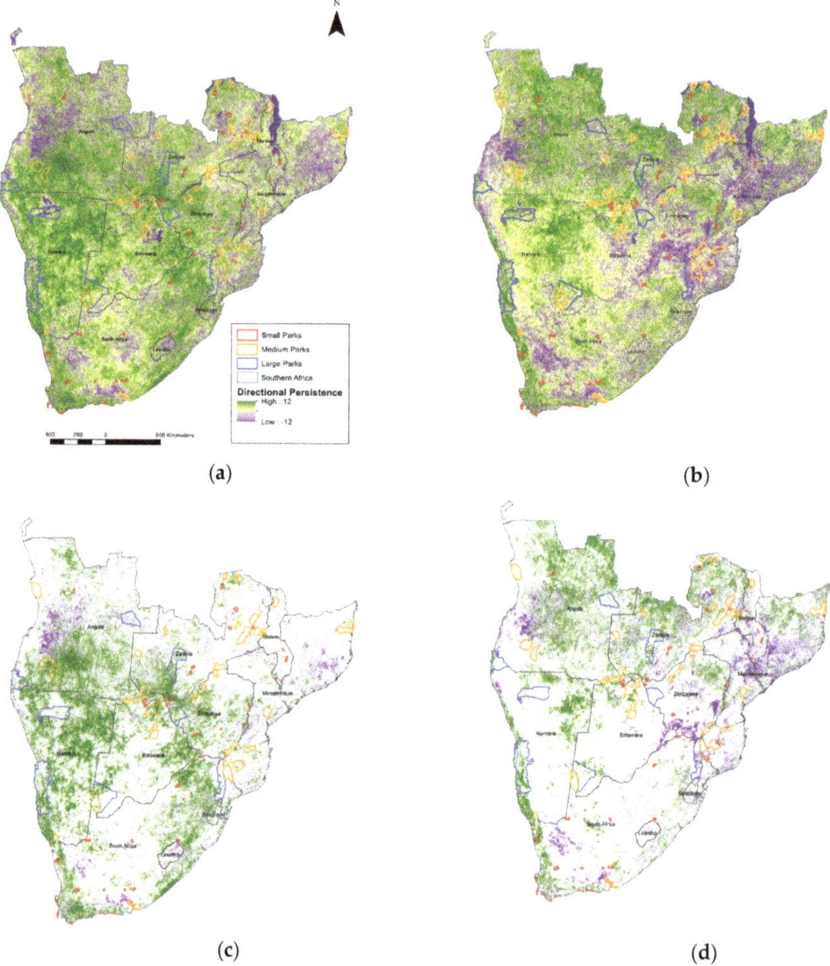

Figure 3. Directional persistence mapped across southern Africa in (**a**) all values in December, January, February; (**b**) all values in March, April, May; (**c**) significant values in December, January, February; and (**d**) significant values in March, April, May.

In MAM there was more neutral persistence across this landscape. In northeastern Namibia and in the Kalahari (western Namibia and South Africa), there were strong positive persistence trends. The corner area where the borders of South Africa, Mozambique, and Zimbabwe meet, as well as in central-northern Mozambique, there were strong negative trends. In southwestern Angola and western central South Africa there were also some negative trends. These trends across seasons may be related to the NDVI factors in the first part of this study, such as precipitation. Further investigation is needed to determine if there is a change in precipitation timing or amount during this study period.

The percentages of significant positive and negative persistence for parks and buffers for each of the three park size classes and two seasons are represented in Figure 4a,b. These figures highlight only significant persistence and do not account for nonsignificant persistence and therefore do not add up to 100%. Across the wet season (Figure 4a,b), in parks and buffer zones, there is an overwhelming significant positive persistence trend. The parks and buffer zones had higher percentages of positive persistence (20%–31%) and lower percentages of negative persistence (3%–4%) in DJF compared to MAM. In MAM, positive persistence ranged from 13% to 26% and negative persistence ranged from 5% to 10%. When comparing parks with buffer zones, in DJF, small and medium parks had higher percentages of positive persistence than buffer zones. Large parks/buffers had similar percentages of positive persistence. In DJF, small and medium parks had lower percentages of negative persistence than buffer zones, while large parks/buffers had similar percentages of negative persistence. When comparing parks with buffer zones, in MAM, small, medium, and large parks had higher percentages of positive persistence than buffer zones. In MAM, small, medium, and large parks had a lower percentage of negative persistence than buffer zones.

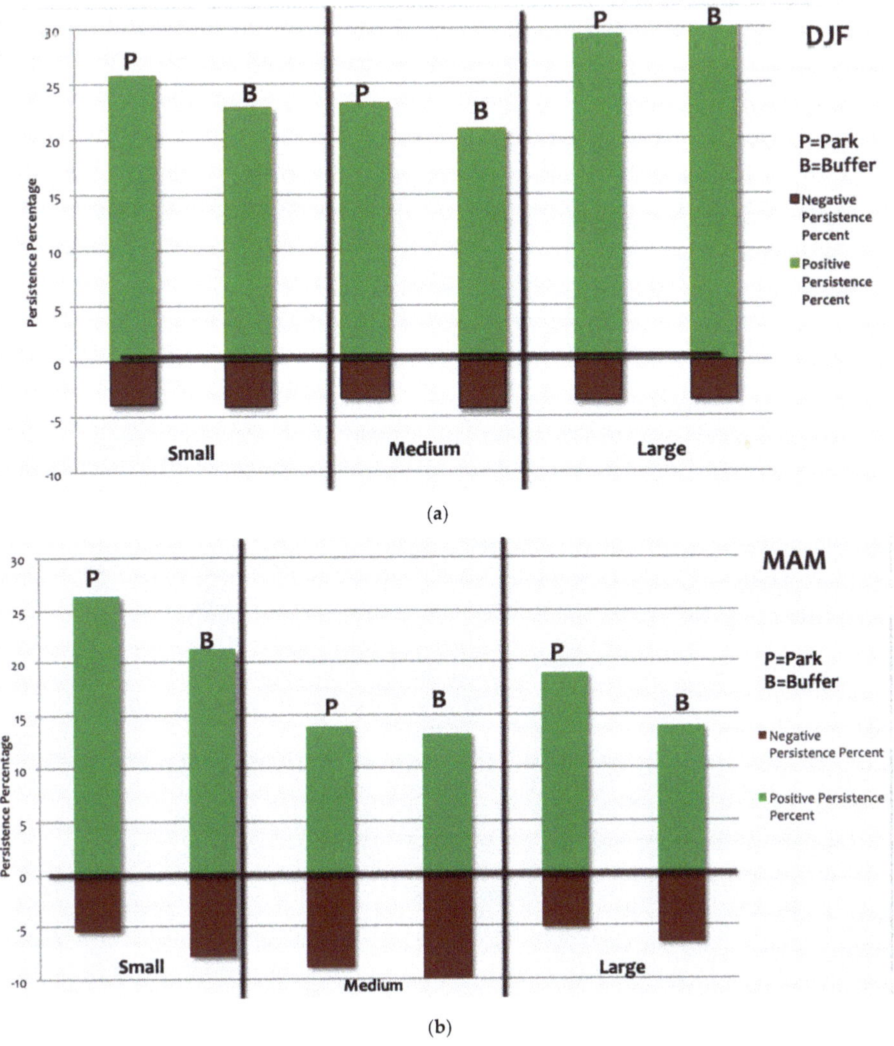

Figure 4. Percentage of significant positive and negative directional persistence values across small, medium, and large parks and buffers in (**a**) December, January, February and (**b**) March, April, May.

4. Discussion and Conclusions

This research supports literature that cites the major drivers in savanna systems as climate, namely precipitation, and humans, namely human population densities that are putting pressure on protected areas [2,32,33]. From this study, we can conclude that across southern African savanna parks and buffers, where the savanna systems are by their very nature water-limited, increases in precipitation lead to higher NDVI. This study supports that precipitation is important in maintaining healthy vegetation, but under an increasingly changing climate, where there are increasing temperatures and more precipitation variability leading to less available moisture, these systems may become significantly altered [7,51]. This can potentially have widespread effects given the extent of drylands globally, and is a reason why such studies, and the development of vegetation persistence metrics, are key.

An unexpected finding of this research was that park size was not a significant factor of determining higher or lower NDVI. The phrase "bigger is better" is often used when describing protected areas because of principles of connectivity in landscape ecology [37], which can be true to a certain threshold, but further nuances in this theory have come about more recently. New information suggests that placement of a protected area is more important than size [38,52,53]. However, "bigger is better" has remained an influence on policy makers. The "Convention on Biological Diversity" and "World Parks Congress" have created initiatives to conserve the maximum area of land as possible, but this may not actually be the best way to conserve biodiversity [52,53]. As DeFries et al. (2007) states, "The relative size and placement of a protected area within the greater ecosystem may be more relevant than absolute size for the ability of a protected area to maintain ecological function" [38]. Therefore, other factors should also be taken into consideration, and one of these is relative location. Protected areas that exist further away from human populations in remote locations and those that have less human use of natural resources tend to function better [38]. Our research supports these previous findings because areas of higher human population density had lower NDVI values across seasons and time periods. Also linked to the relationship between humans and parks is the existence of buffer zones. Buffer zones exist to mitigate the effects of humans on the landscape and wildlife. However, human populations have often been moved out of national parks and into the buffer zones. Therefore, the landscape in buffer zones is often more affected than in parks [54]. This is supported by our findings when comparing vegetation persistence in and out of parks. Vegetation persistence experienced more positive and less negative pixels inside of parks across seasons, time periods, and park sizes when compared to the buffer zones.

This research highlights the importance of the transfrontier conservation areas that have been developed in this century. The principle of transfrontier conservation areas lie in theories of landscape ecology, that being more connected is better [39]. Peace Parks Foundation has begun the effort of creating transfrontier conservation areas across Africa, but several of their pilot sites are in southern Africa and are included in this analysis [43]. Our results further support the validity of setting up these conservation areas across geopolitical borders from the perspective of the positive relationship between the existence of a transfrontier conservation area and higher NDVI values.

This research also supports observed global greening trends in vegetation. Greening of the earth has been observed over the last few decades by NASA and NOAA satellites [24]. One of the largest contributing factors to this greening is CO_2 fertilization, but there are other factors, such as nitrogen levels, land use, temperature, and climate [25–27]. Our research found that the overwhelmingly strong trend in significant directional persistence was positive for much of the region, i.e., that by using NDVI as a measure of vegetation greenness, the years 2005–2016 were greener or had higher NDVI relative to the benchmark years in the earlier part of the 21st century. This greening is not necessarily a positive change biologically. It is also possible that this could be a type of shrub encroachment (correlated local factors, such as increased herbivory, or global factors, such as increased carbon dioxide) or this greening could be from invasive species [55]. This is why it is important for local scale, fieldwork-based studies, in conjunction with these national and macro-scale analyses, to better understand the local landscape dynamics.

The advantage of using the metric of directional persistence is that it provides a robust and statistically driven foundation for more in-depth analysis. Unlike more traditional approaches, this does not just describe the conversion of the system, but gives us more information on vegetation trends over time that could be further broken down into monthly time steps if desired. Directional persistence can be used to understand vegetation dynamics and, from this, we can highlight areas of vulnerability based on patterns of positive and negative persistence over time. Another major advantage of this technique is that each pixel has a statistical significance that the user can specify. This rigorous and innovative approach is an important contribution to the study of vegetated systems globally, as it can be used for any vegetation system. One limitation of this metric is that it quantifies the amount of vegetation rather than type of vegetation. In these savanna systems, determining the grass–shrub–tree components is important for further quantifying degradation. Therefore, these products may be more difficult for local managers to utilize, and more in-depth localized field-based studies should be carried out.

In conclusion, from this study we can determine that, of the variables we included, precipitation, human population, and transfrontier conservation areas are the most important factors of vegetation health in these savanna parks and buffers. For the future, refining rainfall data and including burn area index data as a proxy for fire occurrence may improve these models further. Another variable that is much more difficult to quantify, but may improve models in the future, is that of herbivory on the landscape. Future research will work to incorporate these additional factors of savanna park vegetation.

Across different park sizes and seasons, but especially during rainy seasons, positive persistence is the dominant trend. Positive persistence is higher inside of parks than buffers, and negative persistence tends to be higher inside the buffer zones than in the parks. This suggests that parks are functioning as designed from the perspective that they are important stable features in an increasingly dynamic world. These protected areas are important for conserving biodiversity, which helps to promote tourism and increase the socio-economic activity in this region, which is particularly important to communities who have been largely dependent on subsistence agriculture [56]. While these findings are a regionally important tool, managers should also continue to monitor vegetation at the more localized scale, and their expertise and field-based, site-specific knowledge is of paramount importance. The use of both types of data in concert will provide park managers and conservationists with a very powerful set of analysis tools, both now and in the future.

Author Contributions: Conceptualization, H.H., J.S., and B.C.; methodology, J.S. and H.H.; software, R.K. and E.B.; formal analysis, H.H., R.K., C.M. and E.B.; data curation, R.K., H.H., and E.B.; writing—original draft preparation, H.H. J.S. and C.M; writing—review and editing, J.S., R.K., H.H., E.B., C.M., and B.C. All authors have read and agreed to the published version of the manuscript.

Funding: This research received no external funding.

Acknowledgments: These authors would like to thank the Peace Parks Foundation for the data on transfrontier conservation areas.

Conflicts of Interest: The authors declare no conflict of interest.

Appendix A

Table A1. This table includes the park names and the land cover table, based on majority of pixels, from the Moderate Resolution Imaging Spectroradiometer (MODIS) Collection 5 global land cover [42].

Park	Land Cover (Majority Pixels)
Small Parks	
Agulhas	Deciduous woodland
Augrabies Falls	Sparse grassland

Table A1. *Cont.*

Blue Lagoon	Deciduous shrubland with sparse trees
Bontebok	Deciduous woodland
Cangandala	Deciduous woodland
Cape Peninsula	Background
Chimanimani	Closed deciduous forest
Golden Gate (Qwa Qwa)	Open deciduous shrubland
Golden Gate Highlands	Open deciduous shrubland
Isangano	Deciduous woodland
Karoo	Open grassland with sparse shrubs
Kasanka	Deciduous woodland
Kazuma Pan	Croplands (>5 background %)
Knysna	Background
Lake Malawi	Deciduous woodland
Lengwe	Deciduous shrubland with sparse trees
Liwonde	Deciduous shrubland with sparse trees
Lochinvar	Deciduous shrubland with sparse trees
Luambe	Deciduous shrubland with sparse trees
Lusenga Plain	Closed deciduous forest
Mamili	Open deciduous shrubland
Mapungubwe	Open deciduous shrubland
Marakele	Deciduous woodland
Matobo (Matopos)	Deciduous woodland
Mosi-Oa-Tunya	Croplands (>5 background %)
Mountain Zebra	Open grassland with sparse shrubs
Mountain Zebra/Karoo	Open deciduous shrubland
Mudumu	Open deciduous shrubland
Mushandike	Deciduous woodland
Namaqua	Sparse grassland
Nyanga	Closed deciduous forest
Nyika (Zambia)	Closed deciduous forest
Table Mountain	Deciduous woodland
Tankwa	Sparse grassland
Tankwa Karoo	Sparse grassland
Tsitsikamma	Background
Vaalbos	Open grassland with sparse shrubs
Victoria Falls	Croplands (>5 background %)
West Coast	Open deciduous shrubland
Wilderness	Closed deciduous forest
Zambezi	Deciduous woodland

Table A1. *Cont.*

Medium Parks	
Addo Elephant	Closed evergreen lowland forest
Banhine	Croplands (>5 background %)
Bikuar	Deciduous woodland
Bwabwata	Open deciduous shrubland
Chizarira	Deciduous woodland
Chobe	Open deciduous shrubland
Gonarezhou	Deciduous woodland
Gorongosa	Closed deciduous forest
Kalahari Gemsbok	Sparse grassland
Kasungu	Deciduous woodland
Lavushi Manda	Closed deciduous forest
Limpopo	Closed deciduous forest
Liuwa Plain	Closed grassland
Lower Zambezi	Deciduous shrubland with sparse trees
Lukusuzi	Deciduous shrubland with sparse trees
Makgadikgadi Pans	Open grassland with sparse shrubs
Mana Pools	Croplands (>5 background %)
Matusadona	Deciduous woodland
Mupa	Croplands (>5 background %)
Mweru-Wantipa	Deciduous woodland
North Luangwa	Deciduous shrubland with sparse trees
Nsumbu	Deciduous woodland
Nxai Pan	Open deciduous shrubland
Nyika	Deciduous woodland
Quicama	Open grassland with sparse shrubs
Quirimbas	Deciduous woodland
Richtersveld	Sparse grassland
Sioma Ngwezi	Open deciduous shrubland
South Luangwa	Deciduous shrubland with sparse trees
Zinave	Deciduous woodland
Large parks	
Etosha	Open grassland
Gemsbok	Open grassland
Hwange	Deciduous woodland
Iona	Sandy desert and dunes
Kafue	Deciduous woodland
Kameia	Closed grassland
Kruger	Deciduous woodland
Namib-Naukluft	Stony desert

References

1. Chapin, F.S., III; Chapin, M.C.; Matson, P.A.; Vitousek, P. *Principles of Terrestrial Ecosystem Ecology*; Springer: Berlin/Heidelberg, Germany, 2011; ISBN 1-4419-9504-8.
2. Vogel, M.; Strohbach, M. Monitoring of savanna degradation in Namibia using Landsat TM/ETM$^+$ data. In Proceedings of the Geoscience and Remote Sensing Symposium, 2009 IEEE International, IGARSS 2009, Cape Town, South Africa, 12–17 July 2009; Volume 3, pp. 931–934.
3. Hanan, N.; Lehmann, C. *Tree-Grass Interactions in Savannas: Paradigms, Contradictions, and Conceptual Models*; Taylor and Francis Group: Boca Raton, FL, USA, 2010.
4. Scholes, R.J.; Walker, B.H. *An African Savanna: Synthesis of the Nylsvley Study*; Cambridge University Press: Cambridge, UK, 1993; ISBN 0-521-41971-9.
5. Houghton, E.; Change, I.P. D'experts intergouvernemental sur l'évolution du climat W. In *Climate Change 1995: The Science of Climate Change: Contribution of Working Group I to the Second Assessment Report of the Intergovernmental Panel on Climate Change*; Cambridge University Press: Cambridge, UK, 1996; ISBN 978-0-521-56436-6.
6. Houghton, R.A.; Hackler, J.L.; Lawrence, K.T. The US Carbon Budget: Contributions from Land-Use Change. *Science* **1999**, *285*, 574–578. [CrossRef]
7. Monserud, R.A.; Tchebakova, N.M.; Leemans, R. Global vegetation change predicted by the modified Budyko model. *Clim. Chang.* **1993**, *25*, 59–83. [CrossRef]
8. Ojima, D.S.; Valentine, D.W.; Mosier, A.R.; Parton, W.J.; Schimel, D.S. Effect of land use change on methane oxidation in temperate forest and grassland soils. *Chemosphere* **1993**, *26*, 675–685. [CrossRef]
9. Scholes, R.J.; Archer, S.R. Tree-grass interactions in savannas. *Annu. Rev. Ecol. Syst.* **1997**, *28*, 517–544. [CrossRef]
10. Scholes, R.J.; Hall, D.O. The carbon budget of tropical savannas, woodlands, and grasslands. *SCOPE Sci. Comm. Probl. Environ. Int. Counc. Sci. Unions* **1996**, *56*, 69–100.
11. Andela, N.; Liu, Y.Y.; van Dijk, A.I.J.M.; de Jeu, R.A.M.; McVicar, T.R. Global changes in dryland vegetation dynamics (1988–2008) assessed by satellite remote sensing: Comparing a new passive microwave vegetation density record with reflective greenness data. *Biogeosciences* **2013**, *10*, 6657–6676. [CrossRef]
12. Campo-Bescós, M.A.; Muñoz-Carpena, R.; Southworth, J.; Zhu, L.; Waylen, P.R.; Bunting, E. Combined spatial and temporal effects of environmental controls on long-term monthly NDVI in the southern Africa Savanna. *Remote Sens.* **2013**, *5*, 6513–6538. [CrossRef]
13. Lehmann, C.E.R.; Anderson, T.M.; Sankaran, M.; Higgins, S.I.; Archibald, S.; Hoffmann, W.A.; Hanan, N.P.; Williams, R.J.; Fensham, R.J.; Felfili, J.; et al. Savanna Vegetation-Fire-Climate Relationships Differ among Continents. *Science* **2014**, *343*, 548–552. [CrossRef]
14. Sankaran, M.; Hanan, N.P.; Scholes, R.J.; Ratnam, J.; Augustine, D.J.; Cade, B.S.; Gignoux, J.; Higgins, S.I.; Le Roux, X.; Ludwig, F.; et al. Determinants of woody cover in African savannas. *Nature* **2005**, *438*, 846–849. [CrossRef]
15. Wells, M.; Bradon, K. *People and Parks: Linking Protected Area Management with Local Communities*; World Bank: Washington, DC, USA, 1992.
16. Southworth, J.; Zhu, L.; Bunting, E.; Ryan, S.J.; Herrero, H.; Waylen, P.R.; Hill, M.J. Changes in vegetation persistence across global savanna landscapes, 1982–2010. *J. Land Use Sci.* **2016**, *11*, 7–32. [CrossRef]
17. Carlson, T.N.; Ripley, D.A. On the relation between NDVI, fractional vegetation cover, and leaf area index. *Remote Sens. Environ.* **1997**, *62*, 241–252. [CrossRef]
18. Bégué, A.; Vintrou, E.; Ruelland, D.; Claden, M.; Dessay, N. Can a 25-year trend in Soudano-Sahelian vegetation dynamics be interpreted in terms of land use change? A remote sensing approach. *Glob. Environ. Chang.* **2011**, *21*, 413–420. [CrossRef]
19. Jiang, Z.; Huete, A.R.; Chen, J.; Chen, Y.; Li, J.; Yan, G.; Zhang, X. Analysis of NDVI and scaled difference vegetation index retrievals of vegetation fraction. *Remote Sens. Environ.* **2006**, *101*, 366–378. [CrossRef]
20. Lambin, E.F.; Ehrlich, D. Land-cover changes in sub-saharan Africa (1982–1991): Application of a change index based on remotely sensed surface temperature and vegetation indices at a continental scale. *Remote Sens. Environ.* **1997**, *61*, 181–200. [CrossRef]
21. Tucker, C.J. Red and photographic infrared linear combinations for monitoring vegetation. *Remote Sens. Environ.* **1979**, *8*, 127–150. [CrossRef]

22. Wang, J.; Rich, P.M.; Price, K.P.; Kettle, W.D. Relations between NDVI, Grassland Production, and Crop Yield in the Central Great Plains. *Geocarto Int.* **2005**, *20*, 5–11. [CrossRef]
23. Waylen, P.; Southworth, J.; Gibbes, C.; Tsai, H. Time Series Analysis of Land Cover Change: Developing Statistical Tools to Determine Significance of Land Cover Changes in Persistence Analyses. *Remote Sens.* **2014**, *6*, 4473–4497. [CrossRef]
24. Zhu, Z.; Piao, S.; Myneni, R.B.; Huang, M.; Zeng, Z.; Canadell, J.G.; Ciais, P.; Sitch, S.; Friedlingstein, P.; Arneth, A.; et al. Greening of the Earth and its drivers. *Nat. Clim. Chang.* **2016**, *6*, 791–795. [CrossRef]
25. De Jong, R.; de Bruin, S.; de Wit, A.; Schaepman, M.E.; Dent, D.L. Analysis of monotonic greening and browning trends from global NDVI time-series. *Remote Sens. Environ.* **2011**, *115*, 692–702. [CrossRef]
26. Olsson, L.; Eklundh, L.; Ardö, J. A recent greening of the Sahel—Trends, patterns and potential causes. *J. Arid Environ.* **2005**, *63*, 556–566. [CrossRef]
27. Zhang, K.; Kimball, J.S.; Nemani, R.R.; Running, S.W.; Hong, Y.; Gourley, J.J.; Yu, Z. Vegetation Greening and Climate Change Promote Multidecadal Rises of Global Land Evapotranspiration. *Sci. Rep.* **2015**, *5*, 15956. [CrossRef] [PubMed]
28. Campo-Bescós, M.A.; Muñoz-Carpena, R.; Kaplan, D.A.; Southworth, J.; Zhu, L.; Waylen, P.R. Beyond Precipitation: Physiographic Gradients Dictate the Relative Importance of Environmental Drivers on Savanna Vegetation. *PLoS ONE* **2013**, *8*, e72348. [CrossRef] [PubMed]
29. Staver, A.C.; Archibald, S.; Levin, S. Tree cover in sub-Saharan Africa: Rainfall and fire constrain forest and savanna as alternative stable states. *Ecology* **2011**, *92*, 1063–1072. [CrossRef] [PubMed]
30. Bunting, E.L.; Fullman, T.; Kiker, G.; Southworth, J. Utilization of the SAVANNA model to analyze future patterns of vegetation cover in Kruger National Park under changing climate. *Ecol. Model.* **2016**, *342*, 147–160. [CrossRef]
31. Verweij, R.J.T.; Higgins, S.I.; Bond, W.J.; February, E.C. Water sourcing by trees in a mesic savanna: Responses to severing deep and shallow roots. *Environ. Exp. Bot.* **2011**, *74*, 229–236. [CrossRef]
32. Campbell, A.; Child, G. The impact of man on the environment of Botswana. *Botsw. Notes Rec.* **1971**, *3*, 91–110.
33. Ringrose, S.; Matheson, W.; Wolski, P.; Huntsman-Mapila, P. Vegetation cover trends along the Botswana Kalahari transect. *J. Arid Environ.* **2003**, *54*, 297–317. [CrossRef]
34. Nagy, L.; Grabherr, G. *The Biology of Alpine Habitats*; OUP Oxford: Oxford, UK, 2009; ISBN 978-0-19-856703-5.
35. Campbell, B.M. *The Miombo in Transition: Woodlands and Welfare in Africa*; CIFOR: Bogor, Indonesia, 1996; ISBN 978-979-8764-07-3.
36. Gambiza, J. *A Primer on Savanna Ecology*; Institute of Environmental Studies, University of Zimbabwe: Harare, Zimbabwe, 2001.
37. Newmark, W.D. Insularization of Tanzanian Parks and the Local Extinction of Large Mammals. *Conserv. Biol.* **1996**, *10*, 1549–1556. [CrossRef]
38. DeFries, R.; Hansen, A.; Turner, B.L.; Reid, R.; Liu, J. Land Use Change around Protected Areas: Management to Balance Human Needs and Ecological Function. *Ecol. Appl.* **2007**, *17*, 1031–1038. [CrossRef]
39. Saura, S.; Bertzky, B.; Bastin, L.; Battistella, L.; Mandrici, A.; Dubois, G. Protected area connectivity: Shortfalls in global targets and country-level priorities. *Biol. Conserv.* **2018**, *219*, 53–67. [CrossRef]
40. Protected Planet Protected Planet: World Database on Protected Areas. Available online: https://www.protectedplanet.net/ (accessed on 23 February 2020).
41. Global Historical Climatology Network (GHCN)|National Centers for Environmental Information (NCEI) Formerly Known as National Climatic Data Center (NCDC). Available online: https://www.ncdc.noaa.gov/data-access/land-based-station-data/land-based-datasets/global-historical-climatology-network-ghcn (accessed on 28 February 2020).
42. Friedl, M.A.; Sulla-Menashe, D.; Tan, B.; Schneider, A.; Ramankutty, N.; Sibley, A.; Huang, X. MODIS Collection 5 global land cover: Algorithm refinements and characterization of new datasets. *Remote Sens. Environ.* **2010**, *114*, 168–182. [CrossRef]
43. Home—Peace Parks Foundation. Available online: https://www.peaceparks.org/ (accessed on 24 September 2019).
44. Hocking, R.R. A Biometrics Invited Paper. The Analysis and Selection of Variables in Linear Regression. *Biometrics* **1976**, *32*, 1–49. [CrossRef]

45. Akaike, H. A new look at the statistical model identification. *IEEE Trans. Autom. Control* **1974**, *19*, 716–723. [CrossRef]
46. Measuring Vegetation (NDVI & EVI). Available online: https://earthobservatory.nasa.gov/features/MeasuringVegetation (accessed on 24 September 2019).
47. Thematic Mapping Engine—Thematicmapping.org. Available online: http://thematicmapping.org/engine/ (accessed on 24 September 2019).
48. WorldPop. Available online: https://www.worldpop.org/ (accessed on 24 September 2019).
49. Ensslin, A.; Rutten, G.; Pommer, U.; Zimmermann, R.; Hemp, A.; Fischer, M. Effects of elevation and land use on the biomass of trees, shrubs and herbs at Mount Kilimanjaro. *Ecosphere* **2015**, *6*, 1–15. [CrossRef]
50. Farr, T.G.; Rosen, P.A.; Caro, E.; Crippen, R.; Duren, R.; Hensley, S.; Kobrick, M.; Paller, M.; Rodriguez, E.; Roth, L.; et al. The Shuttle Radar Topography Mission. *Rev. Geophys.* **2007**, *45*. [CrossRef]
51. Vitousek, P.M.; Mooney, H.A.; Lubchenco, J.; Melillo, J.M. Human Domination of Earth's Ecosystems. *Science* **1997**, *277*, 494–499. [CrossRef]
52. Barnes, M.D.; Glew, L.; Wyborn, C.; Craigie, I.D. Prevent perverse outcomes from global protected area policy. *Nat. Ecol. Evol.* **2018**, *2*, 759–762. [CrossRef]
53. Rodrigues, A.S.L.; Andelman, S.J.; Bakarr, M.I.; Boitani, L.; Brooks, T.M.; Cowling, R.M.; Fishpool, L.D.C.; da Fonseca, G.A.B.; Gaston, K.J.; Hoffmann, M.; et al. Effectiveness of the global protected area network in representing species diversity. *Nature* **2004**, *428*, 640–643. [CrossRef]
54. Herrero, H.; Waylen, P.; Southworth, J.; Khatami, R.; Yang, D.; Child, B. A Healthy Park Needs Healthy Vegetation: The Story of Gorongosa National Park in the 21st Century. *Remote Sens.* **2020**, *12*, 476. [CrossRef]
55. Herrero, H.V.; Southworth, J.; Bunting, E. Utilizing Multiple Lines of Evidence to Determine Landscape Degradation within Protected Area Landscapes: A Case Study of Chobe National Park, Botswana from 1982 to 2011. *Remote Sens.* **2016**, *8*, 623. [CrossRef]
56. Child, B. Zimbabwe's CAMPFIRE programme: Using the high value of wildlife recreation to revolutionize natural resource management in communal areas. *Commonw. For. Rev.* **1993**, *72*, 284–296.

Article

Effects of Fence Enclosure on Vegetation Community Characteristics and Productivity of a Degraded Temperate Meadow Steppe in Northern China

Lijun Xu [1], Yingying Nie [1], Baorui Chen [1], Xiaoping Xin [1,*], Guixia Yang [1], Dawei Xu [1] and Liming Ye [2,3,*]

1 Chinese Academy of Agricultural Sciences, Institute of Agricultural Resources and Regional Planning; Hulunbuir Grassland Ecosystem Research Station, Beijing 100081, China; xulijun@caas.cn (L.X.); nieyingying0908@126.com (Y.N.); chenbaorui@caas.cn (B.C.); yangguixia@caas.cn (G.Y.); xudawei@caas.cn (D.X.)

2 Department of Geology, Ghent University, 9000 Gent, Belgium

3 Chinese Academy of Agricultural Sciences, Institute of Agricultural Resources and Regional Planning, Beijing 100081, China

* Correspondence: xinxiaoping@caas.cn (X.X.); liming.ye@ugent.be (L.Y.)

Received: 16 March 2020; Accepted: 22 April 2020; Published: 24 April 2020

Abstract: Species composition and biomass are two important indicators in assessing the effects of restoration measures of degraded grasslands. In this paper, we present a field study on the temporal changes in plant community characteristics, species diversity and biomass production in a degraded temperate meadow steppe in response to an enclosure measure in Hulunbuir in Northern China. Our results showed that the plant community responded positively to the fence enclosure in terms of vegetation coverage, height, above- and belowground biomass. A year-to-year increase in aboveground biomass was observed, and this increase plateaued at the ninth year of the enclosure. Our results also showed that the existing dominant and foundation species gained predominance against other species. The sum of the biomass of these two species was more than doubled after the ninth year of the enclosure. However, belowground biomass only briefly increased until the fifth year of the enclosure and then decreased until the end of the experimental period. Plant diversity, evenness, and richness indices showed similar trends to that of belowground biomass. Overall, we found that the degraded temperate meadow steppe responded significantly positively to the enclosure treatment, but an optimal condition was only reached after approximately 5–7 years of continuous protection, providing a solid use case for grassland conservation and management at regional scales.

Keywords: temperate meadow steppe; grassland degradation; biomass; vegetation; community characteristics; enclosure; restoration; management; policy options

1. Introduction

Grassland provides multiple ecosystem functions in both natural and managed systems. It acts as an ecological barrier for, e.g., farmland systems against environmental infringement. It is also by itself a production base of quality livestock products that provide irreplaceable constituents for food and nutritional security at the global scale [1–4]. However, grassland degradation has extensively been observed in recent years due to over-exploitation [4–8]. The structure, function, and dynamics of grassland ecosystems have received increasing attention in the past few decades in China [8–10]. The grassland acreage in China accounts for 3.5×10^6 km^2, one-fourth of which is distributed in the Inner Mongolia region [4,11,12]. Located in northeastern Inner Mongolia, Hulunbuir possesses the most concentrated and representative temperate meadow steppe in the whole of China, characterized by its native vegetation, rich biodiversity, and diversified landscapes [13,14]. However, the degradation of Hulunbuir grasslands has become more and more evident, and both technological and policy-level countermeasures are urgently needed [7,15–18]. The degradation rate in terms of percent area jumped from less than 10% during the 1980s and 1990s to more than 50% during 2000–2010. Grassland degradation has now become not only a hurdle for regional production but also a threat to the eco-environmental balance in wider areas across the country [14,19].

In a broader context, the grasslands of Inner Mongolia form an important part of the grasslands of Eurasia, with 7.9×10^5 km^2 of natural grassland. Common local practice of grassland management is characterized by a strict functional and spatial delineation between grazing and hay-harvesting [8]. Of the grasslands of Inner Mongolia, 6.4×10^5 km^2 or 81% is used for grazing, while 7.1×10^4 km^2 or 9% is mown for hay, mostly in the Hulunbuir area [12]. As the most important livestock industry region in China, Inner Mongolia owns a multi-year average stocking rate of 1×10^8 standard sheep units (SSU, derived from livestock numbers using a conversion rate of 1 for sheep; 0.8 for goats; and 5 for cattle, horses, and camels [20]). The mutton output of Inner Mongolia accounted for 22% of the national total output in 2018, while 18% of milk consumption in China was produced here [20]. However, high grazing pressure has led to severe degradation and desertification, with decreased forage production and environmental deterioration. Although systematic survey data are not available, remote sensing-assisted estimation suggests that the regional forage production decreased from 1.9×10^8 t in 2005 to 1.3×10^8 t in 2009 [4]. In contrast, actual stocking rate increased from 1.1×10^8 SSU in 2005 to 1.4×10^8 SSU in 2009, resulting in an increase in overgrazing rate from 23% to 112% during the same period [4]. Conversely, continued grassland degradation is expected to harm economic development as well as cause ecosystem instability at the regional to national scales [21]. By the end of the 20th century, 90% of the grassland condition of Inner Mongolia had been degraded to various extents [22], compared to an average of 22% at the national level [7]. This sharp contrast has stimulated growing attention to the protection and restoration of grasslands in recent decades.

Many scientists have engaged in combating grassland degradation. Progress has been made in many areas including rational grazing [7,23,24], enclosure [11,25,26], mowing [12,27,28], rapid planting [7,29,30], etc. Among others, fence enclosure is regarded as a simple yet effective measure [26,31] to increase vegetation coverage [30,32], improve species richness [33,34], raise productivity [35,36], and conserve soil water and nutrients [32,37].

Technically, the duration of the enclosure engagement is an important factor affecting the structure and diversity of plant species in grassland communities [30], especially in the moderately degraded pasture or heavily degraded grassland. An earlier study [38], conducted in a Hulunbuir desertified grassland, showed that longer enclosure duration tended to increase plant diversity and stability. However, some other studies suggested contrary effects of enclosure duration on plant community. For instance, Yan et al. [39] found that longer duration of enclosure reduced plant diversity in the neighboring Horqin desertified grassland. Although the effects of enclosure duration are being disputed, not only based on results from different types of grasslands, but also based on results from the same type of grassland [28], the enclosure itself has been recognized as an economical and effective measure to achieve self-healing and to rebuild ecosystem balance in degraded grasslands [8,26].

On the one hand, beneficial effects can usually be expected from an enclosure treatment. The enclosure creates a physical protection to the grassland from external disturbances, allowing the plants enough time to reproduce, which in turn results in an increased species diversity [40] and a higher chance to restore the plant communities [41]. On the other hand, the enclosure treatment may also give rise to adverse effects. As a matter of fact, enclosure is a human disturbance which eliminates grazing and trampling by large herbivores and blocks the exchange of energy and substances between the enclosed grassland and the external ecosystem [42]. Therefore, finding a balance between beneficial and adverse effects of an enclosure on the community structure and species diversity of grassland ecosystems has become a hotspot of debate. From a practical perspective, finding such a balance also has great significance for the sustainable use of grasslands.

Therefore, we aim to quantitatively determine the optimal duration of the enclosure engagement for the restoration of a typical heavily degraded temperate meadow steppe of the region using field experiments in this research. More specifically, the objectives of this paper are: (1) to determine the effects of a fence enclosure on grassland productivity; (2) to evaluate plant community's responses to the enclosure treatment; (3) to determine the optimal duration of the enclosure treatment; and (4) to formulate technical options and policy recommendations for the restoration and sustainable use of the regional grassland resources. We fulfil these objectives by measuring how long it takes the degraded meadow steppe under the enclosure treatment to reach maximum biomass harvest on a yearly basis, and by analyzing the compositional changes in plant species of the meadow steppe vegetation in relation to the enclosure duration.

2. Materials and Methods

2.1. Site Description

The study was conducted at the Hulunbuir Grassland Ecosystem Observation and Research Station (49°19′35″N, 119°56′52″E) in northeastern Inner Mongolia, China (Figure 1). Temperate continental monsoon climate prevails in the region. Average monthly temperature varies from −48.5 °C in January to 36.2 °C in July, with a mean annual temperature of 1.5 °C. Active accumulated temperature of >10 °C is measured at 1700–2300 degree-days, which corresponds to a frost-free period of approximately 110 days. The mean annual precipitation ranges from 350 mm to 400 mm during 2000–2010, 80% of which falls in June to September [43]. The fenced area under the enclosure treatment is in a typical meadow steppe of the region, whose degradation class was evaluated as heavily degraded [22]. The major foundation species in the plant community is *Stipa baicalensis*, while the dominant species is *Leymus chinensis*, with a range of associated species, including *Cleistogenes squarrosa*, *Melissilus ruthenica*, *Carex duriuscula*, *Ixeris chinensis*, and some other weeds. Kastanozems is the dominant soil type in the region [44,45].

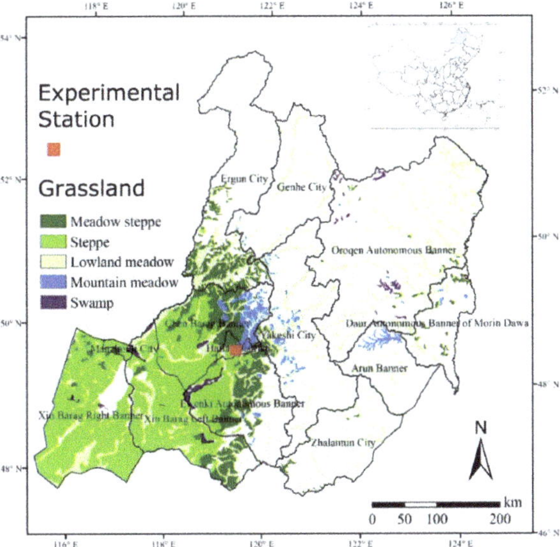

Figure 1. Location of the experimental station in relation to the spatial distribution of grasslands in Hulunbuir. Inset: Geographical location of the study area in China.

2.2. Methods

2.2.1. Experimental Setting

For this study, an experimental plot of 500 m by 320 m was established in Hulunbuir meadow steppe in 2006. A fence was installed on half of the plot, while the other half was left open and used in control experiments. The year 2006 was thus regarded as the first year of the enclosure. A quadrat random sampling design was adopted to determine sample point locations both inside and outside the fenced area (Figure 2). At each sample point, the height, coverage, density, and aboveground and belowground biomass of the plant community within the sampling quadrat were measured. A total of 20 quadrats were randomly chosen, half of which were inside the enclosed plot, and the other half was outside the enclosure. Each quadrat was 1 m × 1 m in size. Plants were sampled annually from 2006 to 2019 in early July when the plant growth was at the maximal level of the year. Samples collected in 2008, 2010, 2012, and 2014 (i.e., the third, fifth, seventh, and ninth year of enclosure; subsequently noted as F3, F5, F7, and F9, respectively) were used in this study.

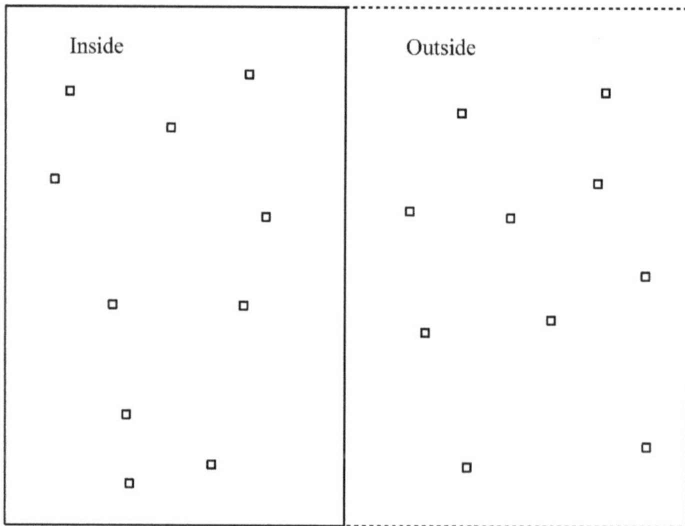

Figure 2. Layout of the quadrats in the experimental plot. A fence is installed along the border of the enclosure represented by the thick line, but not installed along the broken line which is outside the enclosure.

2.2.2. Measurements

The heights of ten randomly selected plants inside each quadrat were measured. The average height was derived from these measurements and used as the height of the plant community. Plant density was measured by counting the total number of plant individuals per quadrat. The area under vegetation cover and the area of bare soil inside each quadrat were visually estimated in-situ by experienced field staff, and the percent vegetative area was derived as the coverage of the vegetation community.

Aboveground biomass was evaluated by collecting and weighing the aboveground parts of the plants in each quadrat per species. Standing plants were cut at the soil surface per species and collected in sample bags. Fallen withered parts above the soil surface were also collected. Plant samples were separated into green parts and dead parts in the laboratory. The fallen withered parts were regarded as dead parts. Collected green and dead parts were weighed to determine fresh weight per sample. Dry weight was also measured after the samples were oven-dried at 85 °C for 12 h.

Belowground biomass was determined by taking soil columns using a root drill (Beijing New Landmark Soil Equipment Co., Ltd, Beijing, China). Three soil columns were taken per quadrat at a depth range of 0–30 cm. The soil columns were water-washed using a 0.5 mm mesh net bag to retain plant roots in the laboratory. The obtained roots were oven-dried at 85 °C for 12 h before weighing.

2.3. Diversity Indices

The relative importance of a plant species in the plant community was evaluated using the plant importance value (PIV), which is given by the following equation [46]:

$$PIV = (RC + RH + RD + RB_a)/4, \tag{1}$$

where RC is the relative coverage of a plant species; RH is the relative height; RD is relative density; and RB_a is the relative aboveground biomass. Here, the values of RC, RH, RD, and RB_a are defined as the percentage of a plant species' value of the coverage, height, density, and aboveground biomass to the total value of all plant species, respectively [46].

Plant diversity was evaluated using four indices, including the Shannon–Wiener diversity index [47], the Simpson diversity index [48], Pielou's evenness index [49] (pp. 89–93), and the Margalef richness index [50]. These indices are given by Equations (2) to (5), respectively.

$$H = -\sum (P_i \cdot \ln P_i), \tag{2}$$

$$D = 1 - \sum P_i^2, \tag{3}$$

$$J = H/\ln S, \tag{4}$$

$$d = (S-1)/\ln N, \tag{5}$$

where P_i is the relative importance value of plant species i, S is the total number of species sampled, and N is the total number of plant individuals sampled.

2.4. Data Analysis

Data were processed and analyzed using Excel (Version 2013, Microsoft Corporation, Redmond, WA, USA) and SPSS (Version 19.0, IBM, NY, USA) packages. A one-way ANOVA was used to test for differences in community structure and diversity indices between the experimental years, while regression modeling was employed to analyze the relationships between above- and belowground biomasses, and between the plant's fallen withered parts and belowground biomass.

3. Results

3.1. Plant Importance Values

The importance values of sampled species both inside and outside the enclosure are summarized in Table 1. The importance values of the dominant species *L. chinensis* and the foundation species *S. baicalensis* were substantially higher inside the enclosure than outside the enclosure. In comparison, the importance values of the species *Artemisia frigida*, *Potentilla bifurca*, and *Carex duriuscula*, which are indicator species of grassland degradation, were lower inside the enclosure than outside the enclosure. The importance values of the other indicator species, *P. bifurca* and *C. duriuscula*, were found to follow similar patterns to species *A. frigida*.

The results of importance values of the species also showed that the importance of the dominant species *L. chinensis* inside the enclosure steadily increased with time and finally reached a maximum value of 30.87% in year F9. In contrast, as the enclosure duration increased, the importance values of the degradation-indicator species *C. duriuscula* and *A. frigida* tended to decrease throughout the enclosure duration, reaching their minimum levels in year F9.

Table 1. The importance values of major plant species inside and outside the enclosure, %.

Species	Enclosure Year							
	F3		F5		F7		F9	
	Inside	Outside	Inside	Outside	Inside	Outside	Inside	Outside
Leymus chinensis	14.80	12.38	20.40	10.39	24.84	9.41	30.87	3.55
Stipa baicalensis	5.44	3.23	6.38	3.69	6.88	1.44	7.39	2.79
Heteropappus altaicuc	1.01	0.17	0.27	0.81	0.76	-	0.48	-
Thalictrum petaloideum	1.82	0.99	0.69	2.70	0.31	0.07	0.78	0.66
Melissilus ruthenica	5.64	2.23	0.49	0.96	0.03	1.25	-	1.34
Bupleurum scorzonerifolium	1.86	0.68	0.73	1.44	0.96	-	0.77	0.47
Artemisia frigida	1.14	1.61	1.07	1.18	0.25	0.45	0.11	1.66
Potentilla bifurca	0.92	1.00	0.27	2.08	1.26	2.47	1.30	1.93
Artemisia laciniate	7.99	9.34	7.86	7.12	8.98	0.68	7.34	5.26
Serratula komarovii	7.21	3.54	4.98	5.02	3.73	0.91	6.18	2.17
Galium verum	1.90	1.25	4.22	1.51	2.40	-	2.10	-
Carex pediformis	11.56	10.70	6.28	9.72	11.12	0.27	12.71	3.00
Adenophora stenanthina	2.54	1.95	4.32	2.13	3.08	0.31	3.90	1.28
Carex duriuscula	2.01	9.67	9.60	10.78	4.59	46.06	0.63	20.89
Palsatilla turczaninovii	6.47	3.99	6.69	11.35	2.82	0.16	2.93	4.69
Cleistogenes squarrosa	1.64	5.27	3.44	2.68	0.45	4.19	0.48	11.31
Achnatherum sibiricum	2.15	-	1.29	1.02	1.84	2.44	5.13	4.39
Iris ventricosa	2.13	1.73	0.29	1.08	2.41	0.30	-	-
Thalictrum squarrosum	1.81	1.02	5.27	3.30	8.01	0.06	3.66	-
Koeleria cristata	-	1.84	-	3.44	1.26	0.19	-	3.54

3.2. Community Height, Coverage, and Density

The variations of plant height, vegetation coverage, and plant density in relation to the enclosure duration are given as bar charts in Figure 3. A comparison between the height of the plants inside and outside the enclosure showed that plants inside the enclosure were significantly higher than plants outside the enclosure. In the third year of the enclosure (F3), for example, plants inside were 189.50% higher than plants outside. After 9 years of the enclosure (F9), plants inside reached a height which was 4.5 times that of the plants outside the enclosure (Figure 3a). Overall, plant height inside the enclosure showed a slight decreasing trend until the fifth year and then a more prominent increasing trend, although statistical significance was not attached. The plant height inside the enclosure maximized in year F9, which was 32.26%, 47.17%, and 13.11% higher than previous years, respectively.

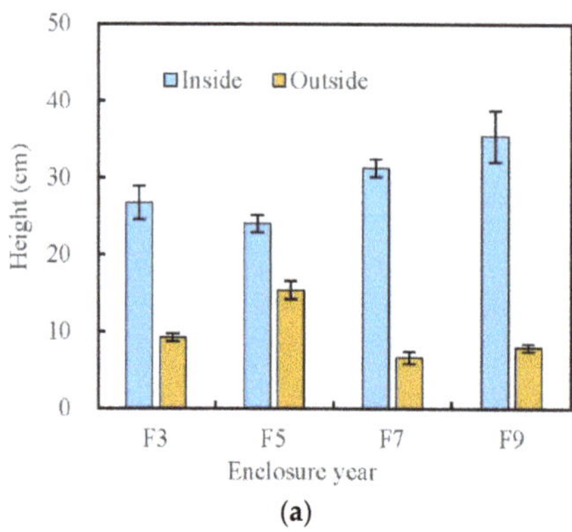

(a)

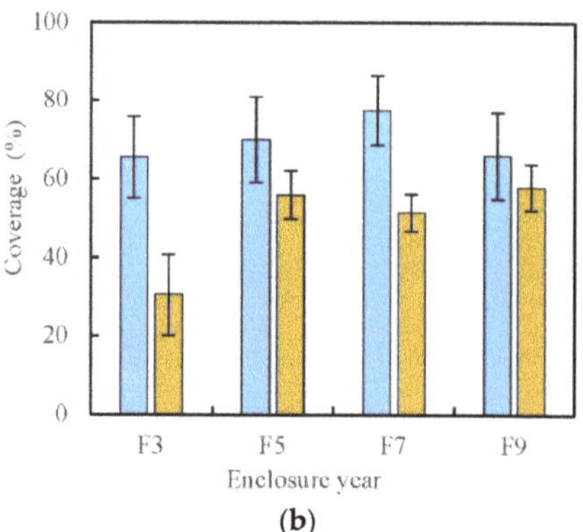

(b)

Figure 3. *Cont.*

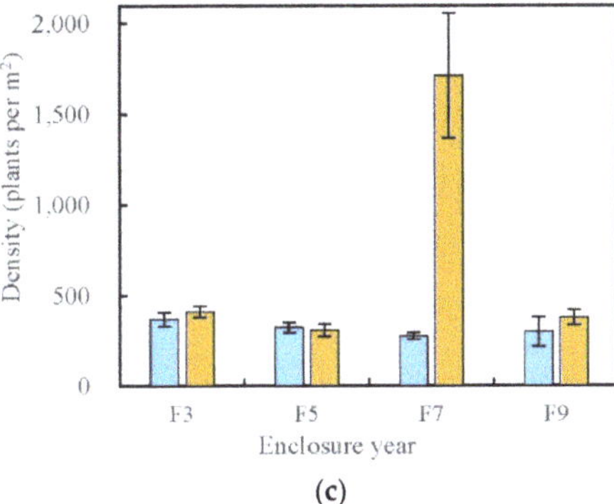

(c)

Figure 3. Variations in plant community characteristics observed both inside and outside the fenced enclosure in the third, fifth, seventh, and ninth years of the enclosure. Observed community characteristics include (**a**) species height in centimeters, (**b**) species' relative ground coverage in percentage, and (**c**) plant density in terms of count of plant individuals per square meter.

The enclosure also had a significant effect on the vegetation cover of the plant community. The plant coverage inside the enclosure was observed to be higher than that outside the enclosure by a margin of 115%, 25%, 50%, and 14% in years F3, F5, F7, and F9, respectively. As the duration of the enclosure increased, the coverage showed a temporal trend of first increasing and then decreasing, and the maximum coverage value was observed in year F7 (Figure 3b).

No significant effect of the enclosure was found on plant density (Figure 3c), although density inside the enclosure was significantly lower than that outside the enclosure in year F7.

3.3. Community Diversity

The variations of the Shannon–Wiener diversity index, the Simpson diversity index, the Pielou's evenness index, and the Margalef richness index values in relation to the enclosure years are shown in Figure 4. The Shannon–Wiener, Simpson, and Margalef indices inside the enclosure were all observed higher than their values outside the enclosure in years F3, F5, and F7. The inside–outside difference of these indices was tested statistically significant in year F7, while significance was not found for the difference in other years. Moreover, only slight inside–outside difference was observed in the Pielou's evenness index values in years F3 and F5 (Figure 4d). In year F9, all the measurements, regardless of diversity, evenness, or richness, showed lower values inside the enclosure than outside the enclosure, albeit not significant.

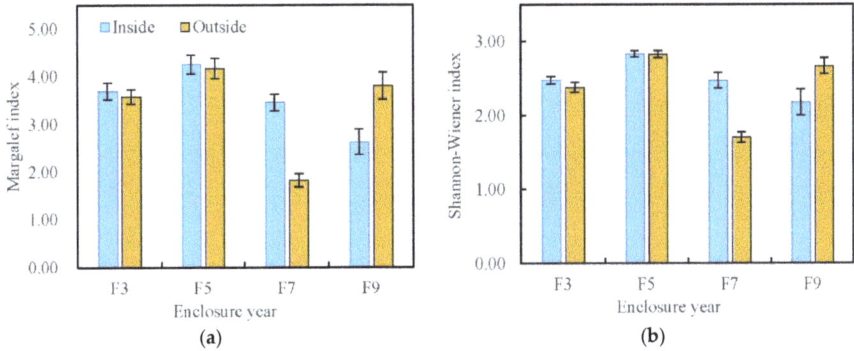

Figure 4. *Cont.*

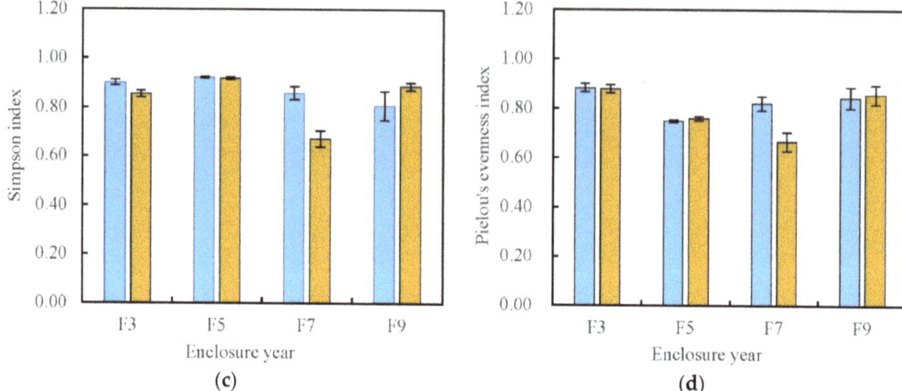

Figure 4. Temporal variations in community diversity of the temperate meadow steppe in Hulunbuir as observed in the plot experiment. Grass species both inside and outside the enclosure were under observation. Multiples indices of plant diversity were derived, including (**a**) Margalef richness index, (**b**) Shannon–Wiener diversity index, (**c**) Simpson diversity index, and (**d**) Pielou's evenness index.

It was notably observed that the Shannon–Wiener and the Simpson diversity indices, together with the Margalef richness index, showed a synchronous pattern whereby these indices tended to increase first and then decrease. All three indices reached a maximum level in year F5 which acted as a turning point. It is interesting to indicate that the Pielou's evenness index showed an opposite trend. The Pielou's index decreased first and then increased, with a turning point also in year F5.

3.4. Biomass Production

The relationship between plant biomass and the enclosure duration is given in Figure 5. The plants inside the enclosure produced higher aboveground biomass than plants outside the enclosure throughout the experimental period (Figure 5a). An inside–outside advantage was achieved at 224.16%, 66.00%, 176.31%, and 309.19% in years F3 through F9, respectively. The general trend was that the aboveground biomass increased with time, reaching the maximum production in year F9. The maximal aboveground biomass in year F9 was higher than in previous years by 65.22%, 46.31%, and 2.56%, respectively, showing a decreasing growth rate.

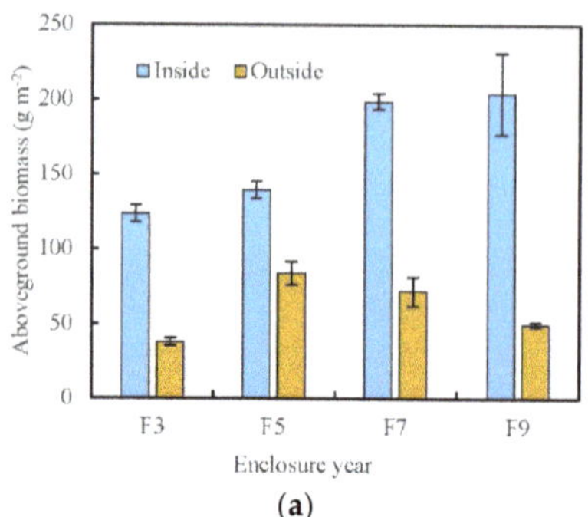

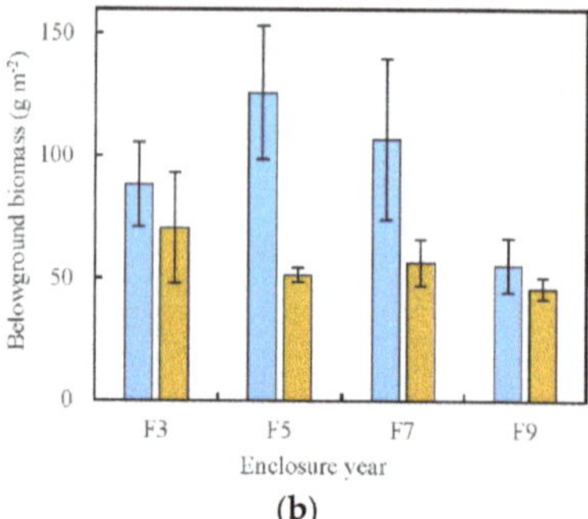

Figure 5. *Cont.*

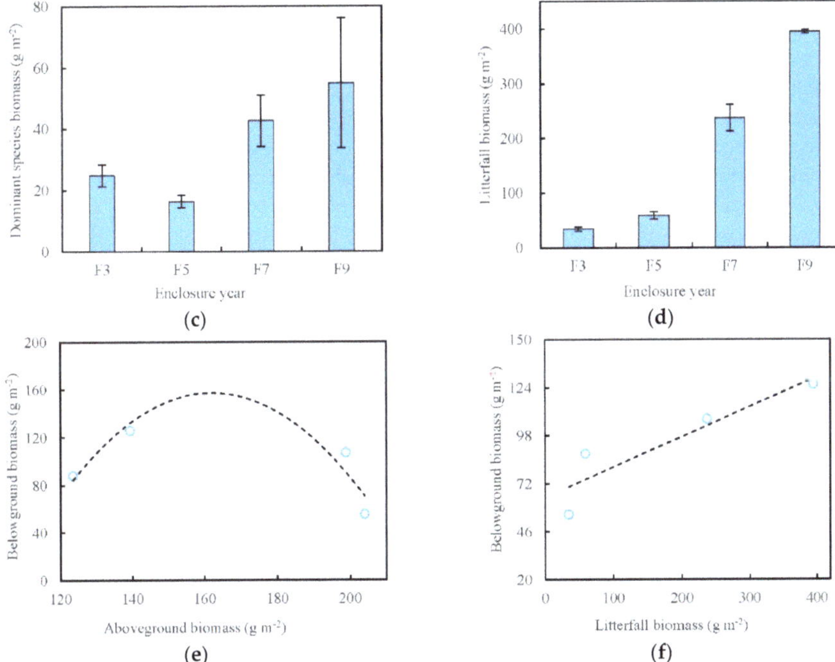

Figure 5. Temporal variations in plant biomass and relationships between biomasses measured above and below the soil surface. Aboveground biomass (**a**) and belowground biomass (**b**) were measured both inside and outside the enclosure, but the aboveground biomass of the dominant species (**c**) and the litterfall biomass (**d**) were only measured inside the enclosure. (**e**) Quadratic relationship between aboveground biomass and belowground biomass: $y = -0.05x^2 + 15.91x - 1131.50$ ($R^2 = 0.79$, $p = 0.001$) and (**f**) linear relationship between litterfall biomass and belowground biomass: $y = 0.12x + 116.65$ ($R^2 = 0.90$, $p = 0.001$).

A similar advantage also existed in belowground biomass under the enclosure treatment (Figure 5b). The belowground biomass measured from soils inside the enclosure was found consistently higher than that outside the enclosure. The inside–outside margin was measured at 25.24%, 143.98%, 89.02%, and 20.47% in years F3 though F9, respectively. Different from the all-time growth pattern in aboveground biomass, the belowground biomass showed a trend of first increasing and then decreasing with a turning point in year F5. Belowground biomass in year F5 was higher than years F3, F7, and F9 by 42.44%, 17.56%, and 126.67%, respectively. Similar increasing trends to that of the aboveground biomass were observed in the biomass of dominant plant species and in the litterfall biomass, both inside the enclosure (Figure 5c,d).

Cause–effect relationships were statistically established from regression analyses involving the belowground biomass versus the aboveground and the litterfall biomasses (Figure 5e,f), showing a quadratic and a linear relationship between the above- and belowground biomass yields and between litterfall and belowground biomass yields, respectively.

4. Discussion

4.1. Plant Community Characteristics

Height, coverage, and density are important characteristics of grassland plant communities [11,13,51,52]. Vegetation coverage and plant density reflect the strength of photosynthetic capacity of green plants, lushness of vegetation cover, and potential of economic traits [9,10]. Our results showed that enclosure caused substantial increases in plant coverage and height. By growing under a condition without cattle grazing and human disturbances, the plants were able to complete a normal life cycle of growth, flowering, setting seeds, etc. Higher quantities of branches, roots and taller plants were consistently observed in the experimental plot. Compared to plants outside the enclosure, the height and coverage of the plant community inside the enclosure were greatly increased (Figure 3a,b). Some tall grasses such as *L. chinensis* are palatable food for cattle, the dominant livestock species in the region. Consequently, the competitive advantage of these tall grasses was effectively suppressed by livestock grazing in open steppe. As such, the shorter species such as *C. duriuscula* obtained more space to grow outside the enclosure, which gave rise to higher density and higher dominance of these species, as confirmed by our results (Figure 3c). Overall, the results obtained from our experimental plot are in line with other works conducted in the region [6,8,12,13,22] or elsewhere under similar conditions [7,17,30,53].

4.2. Plant Diversity

Plant diversity is one of the most critical factors in maintaining the development and productivity of grassland ecosystems [54]. Previous studies on the response of the plant diversity to the enclosure duration have produced mixed results. While some authors (e.g., [55]) suggested that the fence enclosure increased species diversity, others indicated that enclosure had mixed effects on plant diversity [41]. As Cardinale et al. [56] argued, in addition to the enclosure duration, differences in grassland type, habitat, and grazing intensity may all contribute to the differential observational outcomes across the border of the enclosure. In other words, observed differences in community diversity were not necessarily caused by the enclosure treatment alone. In the beginning of our study, species richness and diversity of the enclosed temperate meadow steppe, which had endured severe degradation for years, increased as compared with the open grassland. Even under fence protection, the plant community inside the enclosure did not show a constant upward trend in species richness and diversity. Conversely, it showed a brief increase in species richness and diversity under the enclosure until year F5, followed by a longer period of decrease (Figure 4). There is no doubt that the enclosure prevents livestock from entering and thus creates a beneficial environment with less soil compaction as a direct result of temporary absence of grazing [27,40] and other disturbances [55]. Moreover, enclosure conserves the soil seed bank [36,41], which in turn contributes to the species richness and diversity. However, as the enclosure protection lasts longer in time, the dominant species in the plant community becomes more competitively superior, creating an increasingly tougher environment for the weaker or rarer species due to the so-called compensatory effect [57]. As a result, the species richness and diversity were displaying a downward trend from year F5 to F9 (Figure 4a–c). In contrast, as the enclosure duration increased, the evenness of the plant species in the community first decreased and then increased. It was found that high levels of richness and diversity in the community species led to a high degree of habitat partitioning [34], suggesting that species evenness tended to increase with the enclosure duration (Figure 4d).

4.3. Grassland Productivity

Biomass refers to the total amount of material production, which includes plant parts above the surface and plant roots below the surface. Plant roots play a pivotal role in mediating ecosystem functioning [58]. Moreover, roots bridge the energy and nutrient exchange between above- and belowground ecosystems [59,60]. Biomass is an effective measurement of plant productivity and performance; it is also one of the most basic quantitative characteristics of the grassland ecosystem [61,62]. While aboveground biomass indicates the plant community's ability to accumulate organic matter over a specific duration of time [24], belowground biomass depicts the grassland ecosystem's functioning to input carbon and other critical elements [63]. Our results (Figure 5a–c) showed significantly positive effects of the enclosure on either aboveground, belowground, or dominant species biomass, which was largely in agreement with the results obtained elsewhere [52,53].

Although studies have demonstrated that enclosures could substantially increase aboveground biomass, the duration of grassland enclosures cannot be infinite, given the principles of sustainability for grassland ecosystems [57]. Excessively long duration of enclosure may produce unwanted, adverse effects on the normal growth and development of forage grasses. Unharvested hay piling on soil surface will likely inhibit plant regeneration and seedling growth, thus hindering grassland renovation. It has been reported that proper mowing or grazing tends to encourage the tillering and regeneration of forage grasses and, therefore, to improve the overall quality of grassland [64].

4.4. Optimal Enclosure Duration

Ecological restoration has been defined by the Society of Ecological Restoration International (SER) as the process of assisting the recovery of an ecosystem that has been degraded, damaged, or destroyed [65]. Similarly, the restoration of a degraded grassland can be defined as the process of helping the grassland recover from the degradation status. The recovery can be measured using resilience, stability, and performance attributes of the grassland. Accordingly, the grassland attributes of plant diversity, vegetation coverage, and biomass can be used as proxies of grassland resilience, stability, and performance, respectively [65]. As discussed above, plant diversity, no matter if measured by the Margalef, the Simpson, or the Shannon–Wiener index, reached a maximum value in the fifth year of the enclosure (Figure 4), while the maximum value of the vegetation coverage was reached in the seventh year of the enclosure (Figure 3b). In addition, as an indication of the economic yield of the grassland, aboveground biomass kept an increasing trend at least during the first nine years of the enclosure; however, its growth rate reached a maximum in the seventh year and began to drop from the ninth year onwards (Figure 5a). Moreover, regarded as the energy and nutrient exchange media between the above- and belowground terrestrial ecosystems [58–60], the temporal variation pattern of belowground biomass deserves extra attention. Belowground biomass increased since the beginning of the enclosure and reached a maximum in the fifth year and then began to decrease (Figure 5b). The decreasing trend is so prominent that until the ninth year of the enclosure, the accumulated belowground biomass was less than 50% of the maximum value in the fifth year. Overall, judged on the basis of the variation patterns of these grassland attributes, the ecosystem conditions of the temperate meadow steppe in the study area responded positively to the enclosure treatment. The conditions of this degraded temperate meadow steppe were steadily improved as the enclosure duration lasted. An optimal condition was achieved after approximately 5–7 years under the protection of the enclosure.

5. Conclusions

We conducted a multi-year field experiment to investigate the effects of a fence enclosure on the improvement in quality and wellbeing of a heavily degraded temperate meadow steppe in Hulunbuir in northern China during 2006–2019. Our obtained results clearly show that the fence enclosure is an efficient restoration measure that not only promotes species diversity but also improves grassland productivity. Our obtained results also show that although a degraded grassland can recover steadily under the protection of a fenced enclosure, it takes at least five to seven consecutive years before the grassland vegetation reaches a somewhat stable condition in terms of plant health and productivity. These seemingly simple results have profound technical and policy implications for grassland management and livestock development in the region and beyond.

First of all, our results send a clear warning message to the grassland communities and to the decision-makers that any optimism over the effects of degradation control programs executed recently across the region is likely premature and thus needs to be cautioned. Driven by strong market demand, the grasslands of Inner Mongolia have seen dramatic eco-environmental changes in recent decades. While the livelihood of the grassland communities has largely been improved, grassland degradation and desertification have become more and more prominent. Due to increasing numbers of livestock and changed land use from semi-nomadic systems to a sedentary system, size and productivity of typical steppes in the study region have substantially decreased [66]. In response, the government initiated and implemented a series of grassland restoration programs, such as the Natural Grassland Restoration Program and the Returning Grazing Land to Grassland Program [5]. Meanwhile, a range of policy reforms have been instituted since the beginning of the 21st century, which included the Grassland Eco-compensation Program, the Farmer's Professional Cooperative Law, and so on. Preliminary satellite monitoring data suggested that these programs and institutions produced a significant and positive effect in mitigating degradation rate and accelerating grassland rehabilitation, especially in the pilot counties [18]. However, do these short-term gains necessarily lead to a long-term success? Our data contribute to the scientific understanding of this discussion by proving biophysical evidences on the grassland restoration process. Our experiments show that a degraded grassland has been very responsive to protection measures since the very beginning when grazing disturbances were removed. However, the protection measures need to be maintained and kept in place continuously for a duration of approximately 5–7 years before the grassland ecosystem begins to stabilize. Any improvements from shorter-period intervention measures will be short-lived, too. There is no doubt that the long-term goals of grassland conservation cannot be realized by short-term actions.

Secondly, our experimental evidence also shows that the absence of grazing disturbances is by itself an effective grassland restoration measure. As such, any technical interventions that promote pastural fallowing and mitigate overgrazing should be encouraged. These include, among others, desertification control [21,67], grazing prohibition [5], grazing rotation [68], and artificial grassland plantation [7,30]. Moreover, in order to meet increasing needs of livestock products, the pattern of rearing livestock in captivity, which relies on imported forage as a feed source and lessens the pressure on local grassland, should be further encouraged. In line with the feed demand, expansion of cultivated pasture establishments should be further facilitated [5,18].

Thirdly, policy reforms and institutional changes are needed to provide guidance and support for technical interventions. On the one hand, institutional reforms to reinforce the long-term grassland use rights are needed. Recent policy adjustments to the household responsibility system known as the 'separating three property rights' (STPR) scheme had detrimental effects on grassland protection and conservation. Although STPR was designed to promote grassland transfers [17] and thus to facilitate the economic development in pasture areas, grassland tenants who obtained land use rights under STPR tended to seek short-term gains rather than longer-term sustainable productivity. To cope with this issue, solutions such as financial incentives to stimulate grassland protection [69] and payment for ecosystem services (PES) [70] are preferred. For example, PES can be used as an active intervention to compensate a herder's economic losses caused by putting degraded grasslands into fallowing or by lowering the stocking rate once a carrying capacity threshold is passed [4,71]. On the other hand, trade liberalization of livestock products could help reduce grassland use intensity. Exploring the potential to increase the productivity of the domestic livestock is another way to address this concern, for instance, through the diffusion of techniques for feeding and fattening and by promoting the use of improved breeds [72].

Lastly, we would like to point out that innovations are needed in the current thinking of grassland conservation in China. For example, introduction of greater bottom-up decision-making may be considered at the local level to cope with time-sensitive operations (e.g., rotational grazing) more effectively according to changing weather conditions [73]. Bottom-up decision-making is also needed in situations of, e.g., the prevention of "night-grazing", a rampant illegal action in recent years that breaches grazing prohibition at nighttime when patrol is off duty [74]. Furthermore, involving community-level officials in the co-management of the trading of the grassland use rights, as demonstrated by a Tibetan pasture community in its customary management of grassland and livestock [17], is yet another example of innovative thinking. As a matter of fact, the technical options recommend here are more relevant at the local scale. Without bottom-up decision-making, the feasibility of these options is hardly imaginable. Above all, a balanced livestock–grassland system based on local conditions needs to be established to pave the way for sustainable grassland utilization and livestock production [7].

Author Contributions: Conceptualization, L.X. and X.X.; investigation, L.X. and Y.N.; methodology, L.X. and G.Y.; data curation, B.C., Y.N. and G.Y.; formal analysis, Y.N. and G.Y.; writing—original draft preparation, L.X. and L.Y.; writing—review and editing, L.Y.; visualization, L.Y. and D.X.; supervision, L.Y.; project administration, X.X. and G.Y.; resources, G.Y. and B.C.; funding acquisition, X.X. and L.X. All authors have read and agreed to the published version of the manuscript.

Funding: This research was funded by National Natural Science Foundation of China (No. 41703081), National Key Research and Development Program of China (2016YFC0500603, 2017YFC0503805, 2017YFE0104500), China Agriculture Research System (CARS-34) and National Nonprofit Institute Research Grant of CAAS (912-32).

Acknowledgments: Logistic supports from Ghent University for this paper are kindly acknowledged.

Conflicts of Interest: The authors declare no conflict of interest.

References

1. Hou, X. The development theory and key supporting science and technologies of grasslands and animal husbandry in China. *Pratacul. Sci.* **2015**, *32*, 823–827.
2. Ye, L.; Yang, J.; Verdoodt, A.; Moussadek, R.; Van Ranst, E. China's food security threatened by soil degradation and biofuels production. In Proceedings of the 19th World Congress of Soil Science: Soil Solutions for a Changing World, Brisbane, Australia, 1–6 August 2010; pp. 5–8.
3. Ye, L.-M.; Malingreau, J.-P.; Tang, H.-J.; Van Ranst, E. The breakfast imperative: The changing context of global food security. *J. Integr. Agric.* **2016**, *15*, 1179–1185. [CrossRef]
4. Zhang, Y.J.; Zhang, X.Q.; Wang, X.Y.; Liu, N.; Kan, H.M. Establishing the carrying capacity of the grasslands of China: A review. *Rangel. J.* **2014**, *36*. [CrossRef]
5. Robinson, B.E.; Li, P.; Hou, X. Institutional change in social-ecological systems: The evolution of grassland management in Inner Mongolia. *Glob. Environ. Chang.* **2017**, *47*, 64–75. [CrossRef]
6. Wei, P.; Xu, L.; Pan, X.; Hu, Q.; Li, Q.; Zhang, X.; Shao, C.; Wang, C.; Wang, X. Spatio-temporal variations in vegetation types based on a climatic grassland classification system during the past 30 years in Inner Mongolia, China. *Catena* **2020**, *185*, 104298. [CrossRef]
7. Zhou, W.; Li, J.; Yue, T. *Remote Sensing Monitoring and Evaluation of Degraded Grassland in China: Accounting of Grassland Carbon Source and Carbon Sink*; Springer: Singapore, 2020; p. 138. [CrossRef]
8. Schönbach, P.; Wan, H.; Gierus, M.; Bai, Y.; Müller, K.; Lin, L.; Susenbeth, A.; Taube, F. Grassland responses to grazing: effects of grazing intensity and management system in an Inner Mongolian steppe ecosystem. *Plant Soil* **2011**, *340*, 103–115. [CrossRef]
9. Golodets, C.; Kigel, J.; Sternberg, M. Recovery of plant species composition and ecosystem function after cessation of grazing in a Mediterranean grassland. *Plant Soil* **2009**, *329*, 365–378. [CrossRef]
10. Bai, Y.; Wu, J.; Clark, C.M.; Pan, Q.; Zhang, L.; Chen, S.; Wang, Q.; Han, X.; Wisley, B. Grazing alters ecosystem functioning and C:N:P stoichiometry of grasslands along a regional precipitation gradient. *J. Appl. Ecol.* **2012**, *49*, 1204–1215. [CrossRef]
11. Nie, Y.; Wang, G.; Peng, F.; Du, G. Effects of enclosure on community characteristics in Hunlunbuir meadow steppe. *Chin. J. Grassl.* **2016**, *38*, 87–92.
12. Gu, J.; Wang, M.; Zhao, J. Analysis and evaluation report of grassland resources in Inner Mongolia. *Inn. Mong. Prataculture* **1997**, *1*, 5–12. (In Chinese)
13. Chen, B.; Zhu, L.; Li, G.; Hu, Y.; Xin, X. Analysis of vegetation characteristics under different grazing intensities in Hulunber meadow steppe. *Chin. J. Agric. Resour. Reg. Plan.* **2010**, *31*, 67–71.
14. Lv, S.; Lu, X.; Gao, J. Response of soil fauna to environment degeneration by wind erosion in a Hulunbeir steppe. *Chin. J. Appl. Ecol.* **2007**, *18*, 2055–2060.
15. Ye, L.; Van Ranst, E. Production scenarios and the effect of soil degradation on long-term food security in China. *Glob. Environ. Chang.* **2009**, *19*, 464–481. [CrossRef]
16. Bindraban, P.S.; van der Velde, M.; Ye, L.; van den Berg, M.; Materechera, S.; Kiba, D.I.; Tamene, L.; Ragnarsdóttir, K.V.; Jongschaap, R.; Hoogmoed, M.; et al. Assessing the impact of soil degradation on food production. *Curr. Opin. Environ. Sustain.* **2012**, *4*, 478–488. [CrossRef]

17. Gongbuzeren; Zhuang, M.; Li, W. Market-based grazing land transfers and customary institutions in the management of rangelands: Two case studies on the Qinghai-Tibetan Plateau. *Land Use Policy* **2016**, *57*, 287–295. [CrossRef]
18. Chen, H.; Shao, L.; Zhao, M.; Zhang, X.; Zhang, D. Grassland conservation programs, vegetation rehabilitation and spatial dependency in Inner Mongolia, China. *Land Use Policy* **2017**, *64*, 429–439. [CrossRef]
19. Xia, T.; Wu, W.; Zhou, Q.; Tan, W.; Verburg, P.H.; Yang, P.; Ye, L. Modeling the spatio-temporal changes in land uses and its impacts on ecosystem services in Northeast China over 2000–2050. *J. Geogr. Sci.* **2018**, *28*, 1611–1625. [CrossRef]
20. National Statistical Bureau of China. *China Statistical Yearbook 2019*; China Statistical Press: Beijing, China, 2019; p. 935.
21. Meyer, N. Desertification and restoration of grasslands in Inner Mongolia. *J. For.* **2006**, *104*, 328–331.
22. Zhang, G.G.; Li, X.D.; Kang, Y.M.; Han, G.D.; Hong, M.; Sakurai, K. How to restore the degraded grassland in Inner Mongolia of China? *J. Food Agric. Environ.* **2013**, *11*, 1124–1127.
23. Koerner, S.E.; Collins, S.L. Interactive effects of grazing, drought, and fire on grassland plant communities in North America and South Africa. *Ecology* **2014**, *95*, 98–109. [CrossRef]
24. Oñatibia, G.R.; Aguiar, M.R. Continuous moderate grazing management promotes biomass production in Patagonian arid rangelands. *J. Arid Environ.* **2016**, *125*, 73–79. [CrossRef]
25. Wang, H.; Li, J.; Zhang, Q.; Liu, J.; Yi, B.; Li, Y.; Wang, J.; Di, H. Grazing and enclosure alter the vertical distribution of organic nitrogen pools and bacterial communities in semiarid grassland soils. *Plant Soil* **2019**, *439*, 525–539. [CrossRef]
26. Verdoodt, A.; Mureithi, S.M.; Ye, L.; Van Ranst, E. Chronosequence analysis of two enclosure management strategies in degraded rangeland of semi-arid Kenya. *Agric. Ecosyst. Environ.* **2009**, *129*, 332–339. [CrossRef]
27. Schrama, M.J.J.; Cordlandwehr, V.; Visser, E.J.W.; Elzenga, T.M.; de Vries, Y.; Bakker, J.P. Grassland cutting regimes affect soil properties, and consequently vegetation composition and belowground plant traits. *Plant Soil* **2013**, *366*, 401–413. [CrossRef]
28. Gonzales, E.K.; Clements, D.R. Plant community biomass shifts in response to mowing and fencing in invaded oak meadows with non-native grasses and abundant ungulates. *Restor. Ecol.* **2010**, *18*, 753–761. [CrossRef]
29. Feng, R.; Long, R.; Shang, Z.; Ma, Y.; Dong, S.; Wang, Y. Establishment of Elymus natans improves soil quality of a heavily degraded alpine meadow in Qinghai-Tibetan Plateau, China. *Plant Soil* **2009**, *327*, 403–411. [CrossRef]
30. Shang, Z.H.; Ma, Y.S.; Long, R.J.; Ding, L.M. Effect of fencing, artificial seeding and abandonment on vegetation composition and dynamics of 'black soil land' in the headwaters of the Yangtze and the Yellow Rivers of the Qinghai-Tibetan Plateau. *Land Degrad. Dev.* **2008**, *19*, 554–563. [CrossRef]
31. Yirdaw, E.; Monge, A. Reconsidering what enclosure and exclosure mean in restoration ecology. *Restor. Ecol.* **2018**, *26*, 45–47. [CrossRef]
32. Wu, G.-L.; Du, G.-Z.; Liu, Z.-H.; Thirgood, S. Effect of fencing and grazing on a Kobresia-dominated meadow in the Qinghai-Tibetan Plateau. *Plant Soil* **2008**, *319*, 115–126. [CrossRef]
33. Oba, G.; Vetaas, O.R.; Stenseth, N.C. Relationships between biomass and plant species richness in arid-zone grazing lands. *J. Appl. Ecol.* **2001**, *38*, 836–845. [CrossRef]
34. Petchey, O.L.; Gaston, K.J. Functional diversity (FD), species richness and community composition. *Ecol. Lett.* **2002**, *5*, 402–411. [CrossRef]
35. Osem, Y.; Perevolotsky, A.; Kigel, J. Grazing effect on diversity of annual plant communities in a semi-arid rangeland: interactions with small-scale spatial and temporal variation in primary productivity. *J. Ecol.* **2002**, *90*, 936–946. [CrossRef]
36. Shang, Z.-H.; Deng, B.; Ding, L.-M.; Ren, G.-H.; Xin, G.-S.; Liu, Z.-Y.; Wang, Y.-L.; Long, R.-J. The effects of three years of fencing enclosure on soil seed banks and the relationship with above-ground vegetation of degraded alpine grasslands of the Tibetan plateau. *Plant Soil* **2013**, *364*, 229–244. [CrossRef]
37. Ye, L.; Tang, H.; Van Ranst, E. The role of quantitative land evaluation in food security decision-making in China: The past, present and future. *Bull. Séanc. Acad. R. Sci. Outre-Mer* **2017**, *61*, 415–434.
38. Oliva, G.; Cibils, A.; Borrelli, P.; Humano, G. Stable states in relation to grazing in Patagonia: a 10-year experimental trial. *J. Arid Environ.* **1998**, *40*, 113–131. [CrossRef]

39. Yan, Y.; Tang, H.; Xin, X.; Wang, X. Advances in research on the effects of exclosure on grasslands. *Acta Ecol. Sin.* **2009**, *29*, 5039–5046.
40. Deng, L.; Zhang, Z.; Shangguan, Z. Long-term fencing effects on plant diversity and soil properties in China. *Soil Tillage Res.* **2014**, *137*, 7–15. [CrossRef]
41. Park, K.H.; Qu, Z.Q.; Wan, Q.Q.; Ding, G.D.; Wu, B. Effects of enclosures on vegetation recovery and succession in Hulunbeier steppe, China. *Sci. Technol.* **2013**, *9*, 25–32. [CrossRef]
42. Belsky, A.J. Effects of grazing, competition, disturbance and fire on species composition and diversity in grassland communities. *J. Veg. Sci.* **1992**, *3*, 187–200. [CrossRef]
43. Xu, L.; Xu, X.; Tang, X.; Xin, X.; Ye, L.; Yang, G.; Tang, H.; Lv, S.; Xu, D.; Zhang, Z. Managed grassland alters soil N dynamics and N2O emissions in temperate steppe. *J. Environ. Sci.* **2018**, *66*, 20–30. [CrossRef]
44. Yao, Y.; Ye, L.; Tang, H.; Tang, P.; Wang, D.; Si, H.; Hu, W.; Van Ranst, E. Cropland soil organic matter content change in Northeast China, 1985–2005. *Open Geosci.* **2015**, *7*, 234–243. [CrossRef]
45. Ye, L.; Tang, H.; Zhu, J.; Verdoodt, A.; Van Ranst, E. Spatial patterns and effects of soil organic carbon on grain productivity assessment in China. *Soil Use Manag.* **2008**, *24*, 80–91. [CrossRef]
46. Kent, M. *Vegetation Description and Data Analysis: A Practical Approach*, 2nd ed.; Wiley-Blackwell: Chichester, West Sussex, UK, 2012; p. 414.
47. Spellerberg, I.F.; Fedor, P.J. A tribute to Claude Shannon (1916–2001) and a plea for more rigorous use of species richness, species diversity and the 'Shannon-Wiener' Index. *Glob. Ecol. Biogeogr.* **2003**, *12*, 177–179. [CrossRef]
48. Morris, E.K.; Caruso, T.; Buscot, F.; Fischer, M.; Hancock, C.; Maier, T.S.; Meiners, T.; Müller, C.; Obermaier, E.; Prati, D.; et al. Choosing and using diversity indices: insights for ecological applications from the German Biodiversity Exploratories. *Ecol. Evol.* **2014**, *4*, 3514–3524. [CrossRef]
49. Van Dyke, F. *Conservation Biology: Foundations, Concepts, Applications*, 2nd ed.; Springer: Dordrecht, The Netherlands, 2008; p. 478. [CrossRef]
50. Yeom, D.-J.; Kim, J.H. Comparative evaluation of species diversity indices in the natural deciduous forest of Mt. Jeombong. *Sci. Technol.* **2011**, *7*, 68–74. [CrossRef]
51. Lezama, F.; Baeza, S.; Altesor, A.; Cesa, A.; Chaneton, E.J.; Paruelo, J.M.; De Cáceres, M. Variation of grazing-induced vegetation changes across a large-scale productivity gradient. *J. Veg. Sci.* **2014**, *25*, 8–21. [CrossRef]
52. Jia, Z.; Ma, X.; Xu, C.; Liu, W.; Wei, X.; Lei, S. Effects of short-term enclosure on the vegetation characteristics of slightly degraded alpine meadow in Guinan County. *Acta Pratacul. Sin.* **2019**, *36*, 2766–2774.
53. Mao, S.; Wu, Q.; Zhu, J.; Li, H.; Zhang, F.; Li, Y. Response of the maintain performance in alpine grassland to enclosure on the northern Tibetan Plateau. *Acta Pratacul. Sin.* **2015**, *24*, 21–30.
54. Fırıncıoğlu, H.K.; Seefeldt, S.S.; Şahin, B. The effects of long-term grazing exclosures on range plants in the central Anatolian region of Turkey. *Environ. Manag.* **2007**, *39*, 326–337. [CrossRef] [PubMed]
55. Wang, S.; Fan, J.; Li, Y.; Huang, L. Effects of grazing exclusion on biomass growth and species diversity among various grassland types of the Tibetan Plateau. *Sustainability* **2019**, *11*, 1705. [CrossRef]
56. Cardinale, B.J.; Wright, J.P.; Cadotte, M.W.; Carroll, I.T.; Hector, A.; Srivastava, D.S.; Loreau, M.; Weis, J.J. Impacts of plant diversity on biomass production increase through time because of species complementarity. *Proc. Natl. Acad. Sci. USA* **2007**, *104*, 18123–18128. [CrossRef] [PubMed]
57. Bai, Y.; Han, X.; Wu, J.; Chen, Z.; Li, L. Ecosystem stability and compensatory effects in the Inner Mongolia grassland. *Nature* **2004**, *431*, 181–184. [CrossRef] [PubMed]
58. Erktan, A.; McCormack, M.L.; Roumet, C. Frontiers in root ecology: recent advances and future challenges. *Plant Soil* **2018**, *424*, 1–9. [CrossRef]
59. Ferguson, B.J.; Indrasumunar, A.; Hayashi, S.; Lin, M.-H.; Lin, Y.-H.; Reid, D.E.; Gresshoff, P.M. Molecular analysis of legume nodule development and autoregulation. *J. Integr. Plant Biol.* **2010**, *52*, 61–76. [CrossRef]
60. Limpens, E.; Bisseling, T. Signaling in symbiosis. *Curr. Opin. Plant Biol.* **2003**, *6*, 343–350. [CrossRef]
61. Altesor, A.; Oesterheld, M.; Leoni, E.; Lezama, F.; Rodríguez, C. Effect of grazing on community structure and productivity of a Uruguayan grassland. *Plant Ecol.* **2005**, *179*, 83–91. [CrossRef]
62. Scurlock, J.M.O.; Johnson, K.; Olson, R.J. Estimating net primary productivity from grassland biomass dynamics measurements. *Glob. Chang. Biol.* **2002**, *8*, 736–753. [CrossRef]
63. Dietzel, R.; Liebman, M.; Archontoulis, S. A deeper look at the relationship between root carbon pools and the vertical distribution of the soil carbon pool. *Soil* **2017**, *3*, 139–152. [CrossRef]

64. Antonsen, H.; Olsson, P.A. Relative importance of burning, mowing and species translocation in the restoration of a former boreal hayfield: responses of plant diversity and the microbial community. *J. Appl. Ecol.* **2005**, *42*, 337–347. [CrossRef]
65. Ruiz-Jaen, M.C.; Aide, T.M. Restoration success: How is it being measured? *Restor. Ecol.* **2005**, *13*, 569–577. [CrossRef]
66. Schröder, W.; Schmidt, G.; Schönrock, S. Modelling and mapping of plant phenological stages as bio-meteorological indicators for climate change. *Environ. Sci. Eur.* **2014**, *26*, 5. [CrossRef]
67. Li, J.; Yang, X.; Jin, Y.; Yang, Z.; Huang, W.; Zhao, L.; Gao, T.; Yu, H.; Ma, H.; Qin, Z.; et al. Monitoring and analysis of grassland desertification dynamics using Landsat images in Ningxia, China. *Remote Sens. Environ.* **2013**, *138*, 19–26. [CrossRef]
68. Nie, C.; Li, Y.; Niu, L.; Liu, Y.; Shao, R.; Xu, X.; Tian, Y. Soil respiration and its Q10 response to various grazing systems of a typical steppe in Inner Mongolia, China. *PeerJ* **2019**, *7*, e7112. [CrossRef]
69. Li, A.; Wu, J.; Zhang, X.; Xue, J.; Liu, Z.; Han, X.; Huang, J. China's new rural "separating three property rights" land reform results in grassland degradation: Evidence from Inner Mongolia. *Land Use Policy* **2018**, *71*, 170–182. [CrossRef]
70. Chen, H.L.; Lewison, R.L.; An, L.; Tsai, Y.H.; Stow, D.; Shi, L.; Yang, S. Assessing the effects of payments for ecosystem services programs on forest structure and species biodiversity. *Biodivers. Conserv.* **2020**, in press. [CrossRef]
71. Tang, H.; Ye, L. *Comparative Study on Methodology of Land Production Potential*; China Agricultural Science and Technology Press: Beijing, China, 1997; p. 301.
72. Liu, M.; Dries, L.; Heijman, W.; Zhu, X.; Deng, X.; Huang, J. Land tenure reform and grassland degradation in Inner Mongolia, China. *China Econ. Rev.* **2019**, *55*, 181–198. [CrossRef]
73. Tang, R.; Gavin, M.C. Degradation and re-emergence of the commons: The impacts of government policies on traditional resource management institutions in China. *Environ. Sci. Policy* **2015**, *52*, 89–98. [CrossRef]
74. Dong, S.; Kassam, K.-A.S.; Tourrand, J.F.; Boone, R.B. *Building Resilience of Human-Natural Systems of Pastoralism in the Developing World: Interdisciplinary Perspectives*; Springer: Cham, Switzerland, 2016; p. 298.

Article

A Tale of Grass and Trees: Characterizing Vegetation Change in Payne's Creek National Park, Belize from 1975 to 2019

Luke Blentlinger * and Hannah V. Herrero

Department of Geography, University of Tennessee, 1000 Philip Fulmer Way, Knoxville, TN 37996-0925, USA; hherrero@utk.edu

* Correspondence: lblentli@vols.utk.edu

Received: 27 May 2020; Accepted: 22 June 2020; Published: 25 June 2020

Abstract: The lowland savannas of Belize are important areas to conserve for their biodiversity. This study takes place in Payne's Creek National Park (PCNP) in the southern coastal plain of Belize. PCNP protects diverse terrestrial and coastal ecosystems, unique physical features, and wildlife. A Support Vector Machine (SVM) classification technique was used to classify the heterogeneous landscape of PCNP to characterize woody and non-woody conversion in a time-series of remotely sensed data from 1975, 1993, 2011 and 2019. Results indicate that the SVM classifier performs well in this small savanna landscape (average overall accuracy of 91.9%) with input variables of raw Landsat imagery, the Normalized Difference Vegetation Index (NDVI), elevation, and soil type. Our change trajectory analysis shows that PCNP is a relatively stable landscape, but with certain areas that are prone to multiple conversions in the time-series. Woody vegetation mostly occurs in areas with variable slopes and riparian zones with increased nutrient availability. This study does not show extensive woody conversion in PCNP, contrary to widespread woody encroachment that is occurring in savannas on other continents. These high-performing SVM classification maps and future studies will be an important resource of information on Central American savanna vegetation dynamics for savanna scientists and land managers that use adaptive management for ecosystem preservation.

Keywords: remote sensing; neotropical savannas; woody conversion; support vector machine classification; land cover change; protected areas; Payne's Creek National Park

---

1. Introduction

Savannas are mixed woodland–grassland ecosystems with continuous grass cover and variable tree density that can border closed tropical forests. The tropical savanna climate, characterized by its distinct wet and dry seasons, and corresponding ecoregions broadly occupies the space between the equatorial tropical rainforests and subtropical deserts [1,2]. Covering about 20% of the Earth's land surface, savannas account for 30% of terrestrial net primary production [3]. The composition of savanna vegetation is some combination of grassland and sporadic woodland in a patchwork mosaic. The factors that determine boundaries between savanna and forest have generated debate [4]. It has been suggested that the factors that control savanna distribution is variable between continents [3] and that the globally ubiquitous woody encroachment occurring in savannas also varies at continental and regional scales [5]. The determinants of savanna vegetation also vary at regional and local scales [6]. Climate is the primary driver of the distribution of savanna vegetation, especially at continental and regional scales, while topography, soil, fire, and herbivory vary in influence regionally to locally [6].

In some global remote sensing studies, tropical savannas are broadly divided into those that occur in South America, Africa and Australia [3,7,8], possibly because they occupy large areas and are widely classified as savanna (or associated ecoregions) in publicly available datasets, such as the World

Wildlife Fund ecoregions [9]. Global studies provide valuable information on the broad distribution, drivers, and health of savanna vegetation on a continental scale. While vast savannas are widely studied and are included in global studies, smaller pockets of savanna have received less attention. There is a lack of understanding of the fine-scale drivers of the heterogeneous nature of savanna vegetation [10]. Vegetation and soil studies in the expansive Brazilian Cerrado have concluded that fine-scale interactions between vegetation and available water [10], and vegetation and soils [11], are key to understanding neotropical savanna vegetation dynamics. Since knowledge of local dynamics is crucial for understanding the larger savanna landscape, investigating vegetation drivers in smaller pockets of savanna can also contribute to the larger understanding of savannas. Smaller savannas are also key components in the preservation of biodiversity and contributions to the economy via tourism [12]. The exclusion of relatively small savannas from global or continental remote sensing studies can be partly attributed to the use of sensors with large spatial resolutions (e.g., Advanced Very High Resolution Radiometer with 4 km pixel size) that contain much intra-pixel variability. For small study areas that focus on localized savanna vegetation dynamics, finer spatial resolution imagery is necessary to sufficiently characterize the area and the heterogeneous vegetation gradients that are characteristic of savanna landscapes. Even when using imagery from sensors with finer spatial resolution, defining the limits of vegetation in savannas is challenging due to the randomness and spatial heterogeneity of the landscape [13].

Another challenge is that savanna vegetation exists in varying compositions and gradients, and one of the most common ways to assess savanna vegetation with remote sensing is to create discrete land cover classes using an unsupervised or supervised technique [14]. These techniques use a set of variables (inputs) to break each pixel in an image into distinct classes (output). Discrete classifications are simple to understand but prevent variation within classes, which, especially for savannas, can result in an inaccurate representation of the continuous and varied transitions between vegetation. The complicated nature of applying classification techniques to remotely sensed imagery in small savanna landscapes suggests that statistically robust methods should be used. Many common supervised classification techniques, such as Maximum Likelihood, function on the assumption that the statistics for each class are normally distributed and then calculate the probability that a pixel belongs to a specific class [15,16]. Since savanna landscapes often do not fit this definition of normality, such a classification technique is inappropriate. One method to improve supervised classification performance in savanna landscapes is to use non-parametric machine-learning algorithms that do not make assumptions about the distribution of the data, such as Support Vector Machine (SVM) or Random Forest [17,18]. Both SVM and Random Forest have been used to generate high-performing vegetation classifications in some African savanna landscapes [12,19,20].

Savannas in the neotropics cover approximately two million km^2, or 40% of all land area in the ecozone [21,22]. Of these, the South American savannas (e.g., Brazilian "Cerrado" and nearby Bolivian "Llanos", Venezuelan-Colombian "Llanos", etc.) have gained the most attention due to their vastness, high biodiversity and biomass, and global importance in the carbon cycle [21]. While these large savanna expanses are undoubtedly important areas of study, smaller landscapes combined with other information can allow researchers to better understand local drivers of vegetation, such as the influence of topography, soil properties, and disturbances. There are isolated savannas in Central America and the Caribbean that occur above and below 1000 m elevation [23]. While all savannas are in part determined by climatic factors, these savannas are heavily influenced by the underlying physical geography. The upland savannas are characterized by weathered soils, fluvial downcutting and riparian forests, and cooler-weather grasses. The lowland savannas are generally located in flat coastal plains and are created by the interaction between climate, topography, drainage, and soils. Both the upland and lowland savannas are also characterized by disturbance in the form of fire and hurricanes [23]. These large areas, underlain by nutrient-poor soil, have resulted in a general lack of human land use since the times of the Maya.

More recently, some of these savannas have been designated as protected areas for reasons ranging from biodiversity to economics. National parks and other protected areas are ideal for analyzing localized vegetation dynamics for a variety of reasons. Protected areas theoretically act as a static feature in dynamic human-influenced landscapes, and represent a more natural state of being in a given biome [24]. Garbulsky and Paruelo [25] used remote sensing techniques in Argentinian protected areas to derive the "baseline/reference" conditions of various ecosystems as to better assess the impacts of land use and global change. Protected areas act as areas of preservation for biodiversity, and this attracts tourism to boost the economy [12]. In addition, protected areas are often small enough that moderate or fine spatial resolution remotely sensed data is appropriate for an analysis of fine-scale vegetation drivers. Nevertheless, even when coarse imagery is most appropriate, imagery with finer spatial resolution may be able to supplement the analysis. Despite the name, many protected areas are heavily influenced by human involvement or have been in the past. Josefsson et al. [26] used paleoecological and archaeological evidence of long-term human land use in boreal forest protected areas to denounce the notion that purportedly "pristine" forests can be used as ecological references. Hence, protected areas may be an indicator of how local, regional, and global human-environment interactions have affected floral and faunal life in the past and present. This reinforces the need to consider multiple lines of evidence to ascertain the drivers of vegetation change in any landscape [5]. Finally, protected areas are often managed by governments or non-governmental organizations, which means that there are often concerted efforts to maintain and protect the vegetation within a protected area. For example, April Sahara et al. [27] used dendrochronological and remote sensing techniques to characterize woody encroachment over the past 150 years in a pine savanna in Redwood National Park, California. Using their findings, they created models to predict the spatiotemporal characteristics of future woody encroachment in the park [27].

This research was carried out in Payne's Creek National Park (PCNP) located in the southern coastal plain of Belize. The purpose of this study is to use remotely sensed data to quantify vegetation land cover change from 1975 to 2019, specifically to determine if there has been any woody conversion, as seen in other savanna landscapes globally. This study addresses the following questions: (1) Can a SVM accurately classify the complex savanna landscape of PCNP over time? (2) What is the overall pattern of woody and non-woody conversion in PCNP from 1975 to 2019 using the SVM classifications? (3) What are the fine-scale variables linked to vegetation distribution in PCNP?

2. Materials and Methods

2.1. Study Area

Payne's Creek National Park (PCNP) in the Toledo District of southern Belize covers approximately 152 km^2 of low elevation land that borders the Caribbean Sea (Figure 1). The park is located at approximately 16.3° N 88.6° W in Belize's southern coastal plain. Belize has a subtropical to tropical climate with mean monthly minimum/maximum temperatures from 16/28 °C in winter to 24/33 °C in summer [28]. The Köeppen climate classification of most of Belize, including PCNP, is "tropical monsoon" [29], which is characterized by a distinct wet and dry season. The majority of rainfall falls from June to November in Belize [29]. For this study, the dry season is considered as January–May and the wet season is June–December. The dry season average precipitation in PCNP is approximately 498 mm and 2208 mm in the wet season. The minimum/maximum precipitation in PCNP in the dry season was 269 mm/790 mm (1987/1990) and 1647 mm/2903 mm (2004/1979) in the wet season [30–32]. Mean annual rainfall ranges from 1524 mm in the north and 4064 mm in the south. This variability occurs due to orographic lifting over steep slopes and the seasonal migration of the Inter-Tropical Convergence Zone [29]. Belize contains upland savannas known as the Mountain Pine Ridge and lowland savannas of the northern and southern coastal plains. This study focuses on Belize's lowland savannas.

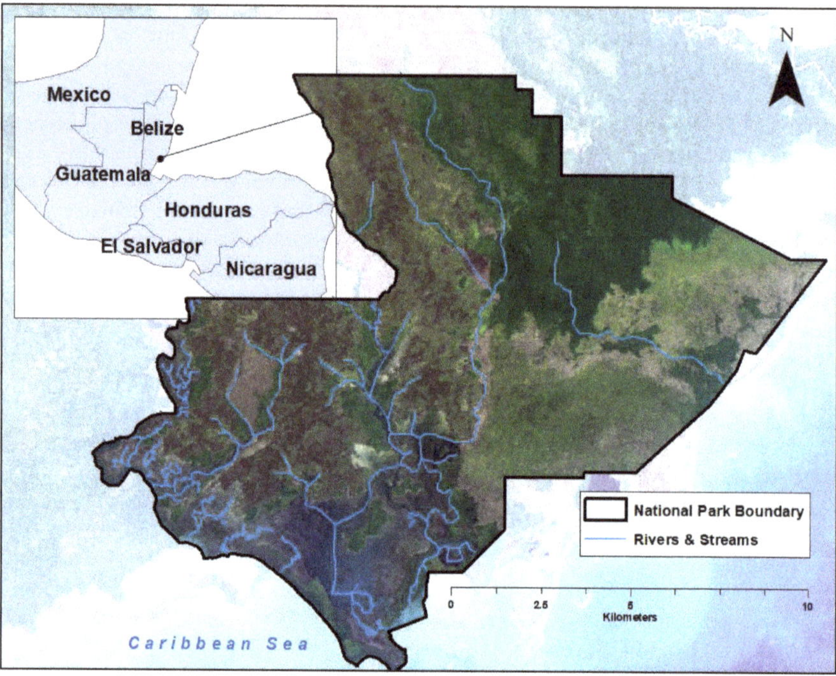

Figure 1. Map of Payne's Creek National Park (PCNP), showing nearby countries in Central America. Image of PCNP is a true color composite from Landsat 5 TM taken in March 2011. Shapefile of Central America from Natural Earth. Shapefile of Belize and its rivers from Biodiversity and Environmental Resource Data System of Belize. Shapefile of Payne's Creek National Park from Protected Planet.

The lowland savannas of Belize are located between dense broadleaf rainforests inland and mangrove swamps seaward in the relatively flat, poorly drained coastal plain [33]. Vegetation in these savannas are a mosaic of grassland and pine or oak trees, and pockets of broadleaf forest [23,28,34]. Both upland and lowland savannas are in areas with weathered, acidic and nutrient-poor soils, but the upland savannas are well-drained while those in the lowlands frequently experience flooding and drying with the wet and dry seasons, respectively [35]. The lowland area is underlain by the Yucatan limestone platform, which was shallow sea during the Pliocene (2–13 Myr BP). Subsurface clay sits atop the limestone platform, resulting in widespread poor drainage in the lowlands. Sands and gravels from the nearby Maya Mountains eroded and accumulated on the clay and formed the current lowland coastal plain [35]. Savannas in the low-lying area usually occur on nutrient-poor soils underlain by the sands and gravels [28]. Undulations in the lowland topography result in improved drainage and the dominance of woody vegetation, such as pine and oak. Dense broadleaf forests occur along rivers in the area due to increased nutrients in alluvial deposits originating in the Maya Mountains [28]. At an even finer scale, disturbance feedbacks in the form of fire and hurricanes control the boundaries between grasses and trees. This is partly because the dominant pine species, *Pinus caribaea*, most readily regenerates on sunlit mineral soils. These soils become available after fire from the removal of grass cover and organic matter, or because old trees are killed and stand density is reduced [36]. Similarly, strong winds from hurricanes can reduce stand density and allow sunlight to penetrate to the surface [36]. The Toledo Institute for Development and Environment (TIDE) manages PCNP and uses fire management practices in the fire-adapted tropical pine savanna and grassland ecosystems. While the pine savanna is fire adapted, too many fires would eventually lead to a grassland without pine by scorching saplings attempting to regenerate. Too few fires results in a thick herbaceous layer and succession towards broadleaf forest. One result of less pine is habitat loss for the endangered yellow-headed parrot, *Amazona oratrix*, which uses pine trees to nest [37]. Since the pine savanna and yellow-headed parrot are both endangered and reliant on fire, TIDE places a high priority on fire management in PCNP.

According to the ecosystem classifications created by other researchers [33,38], PCNP contains open savanna, dense tree savanna, forest inclusion (within open savanna), forest, mangrove and littoral swamp, and wetland. Open savanna is dominated by grasses and sedges with semi-open to very open areas of pine, oak, palmetto and craboo. Dense tree savanna is the transitional area from forest or forest inclusion to open savanna. Elevation broadly distributes the wetland in the lowest lying areas and forest in the higher elevations (<35 m), with savanna in between. The mangrove and littoral swamps are on the coast and also extend inland, beyond the mouth of Payne's Creek and other streams.

2.2. Climate Data

Climate data was used to avoid anomalies and assess the viability of carrying out dry-season vegetation classification and change trajectory analysis in the months from which the imagery was collected. We used two sources of historical climate data to encompass the entire time period. The Climate Hazards Group InfraRed Precipitation with Station Data (CHIRPS) from the University of California Santa Barbara is a quasi-global rainfall dataset that spans 50° S–50° N across all longitudes. CHIRPS uses 0.05° resolution satellite imagery and in-situ station data to produce a gridded, monthly dataset the spans from 1981 to near-present [32]. Since the remote sensing analysis contains the earliest usable Landsat imagery of PCNP from 1975, the CHIRPS data does not encompass the earliest part of the study period. WorldClim is another gridded dataset for historical climate data. Specifically, the CRU-TS 4.03 [31] downscaled with WorldClim 2.1 [30] is a monthly climate dataset for global land areas at approximately 21 km spatial resolution. Since this dataset provides climate data that extends back to 1960, it was used to extend our climate data back to 1970. Even though the CHIRPS and WorldClim data are at different spatial resolutions, both were used to assess the climate in PCNP. Based on the 49 years of combined CHIRPS and WorldClim precipitation data, the dry season (January–May) average precipitation in PCNP is approximately 498 mm and 2208 mm in the wet season [30–32]. Figure 2, which shows the yearly cumulative precipitation Z-scores for the CHIRPS and WorldClim data from 1970 to 2018, was used to assess which years were drier or wetter than average precipitation in PCNP to ensure that anomalous years of imagery were not chosen for analysis.

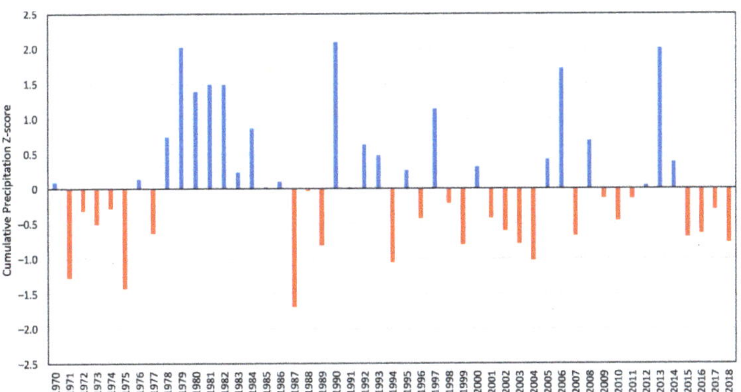

Figure 2. Total annual rainfall Z-scores for PCNP from 1981 to 2019. Rainfall data from WorldClim and CHIRPS.

2.3. Data and Image Analysis

2.3.1. Remotely Sensed Data

This study used remotely sensed Landsat Multispectral Scanner (MSS), Thematic Mapper (TM), and Operational Land Imager/Thermal Infrared Sensor (OLI/TIRS) imagery to create land cover classification and change trajectories and carry out Normalized Difference Vegetation Index (NDVI) analysis on freely available imagery from four years in the dry season: 1975, 1993, 2011 and 2019. Landsat imagery was chosen for this analysis for its relatively fine spatial resolution and its long-term database of publicly available satellite imagery. All Landsat imagery was obtained from the United States Geological Survey Earth Explorer and came atmospherically and geometrically corrected [39]. The images from March 1975 and March 2011 are completely cloud-free, while those from March 1993 and January 2019 have some clouds and cloud shadows, which were masked out. The MSS data (bands 1–4), obtained by Landsat-2 in 1975 at 80 m spatial resolution, were resampled to 30 m pixel size to align with the resolution of the TM (bands 1–7) and OLI/TIRS (bands 1–8, 10, 11) imagery from Landsats 4, 5 and 8 (1993, 2011 and 2019). All raw image bands (except OLI band 9) were subset to a shapefile of PCNP obtained from Protected Planet and were used in the classification analysis. Additionally, the three other layers outlined in Section 2.3.2 below were created or obtained to be used as variables in the SVM classification.

2.3.2. Other Data: Normalized Difference Vegetation Index (NDVI), Digital Elevation Model (DEM), and Soil Type

NDVI is a commonly used vegetation index that has a variety of applications, but its inherent use is to quantify and discriminate vegetation characteristics [40]. The calculation for *NDVI* is:

$$NDVI = \frac{NIR - Red}{NIR + Red}$$

where *NIR* is near infrared reflectance and *Red* is red reflectance in the electromagnetic spectrum. *NDVI* is an indicator of vegetation productivity and degradation [40,41]. Using band math tools in the Environment for Visualizing Images (ENVI) 5.5 software, four NDVI maps, one for each image in the time series, was created from the *NIR* and red bands to be used as an input variable (Figure 3).

As was previously stated, edaphic factors are key to understanding the distribution of woody and non-woody vegetation in the lowland savannas of Belize. Nutrient-poor soils in the Belize lowlands stem from inundations in the wet season due to poor drainage from subsurface clay and intense desiccation in the dry season [35]. Higher elevations and variable slopes can result in locally improved drainage and the opportunity for soil development and woody vegetation establishment. To encompass the importance of topography in the classification, a 30 m digital elevation model (DEM) from the USGS Shuttle Radar Topography Mission (SRTM) was included (Figure 3). The DEM was acquired from the USGS Earth Explorer website and the study area was subset in ArcMap 10.7.1 using the PCNP shapefile.

While the subsurface clay in PCNP affects vegetation through seasonal inundations, the soil types that overlay the clay also affect vegetation distribution. Alluvial deposits along rivers and streams, for example, are nutrient rich and promote woody growth compared to the Maya Mountain outwash that underly the grasslands [28]. To incorporate the importance of soil type in the classification, a Belize soils map created by the Selva Maya consortium (Figure 3) was included as a variable [42].

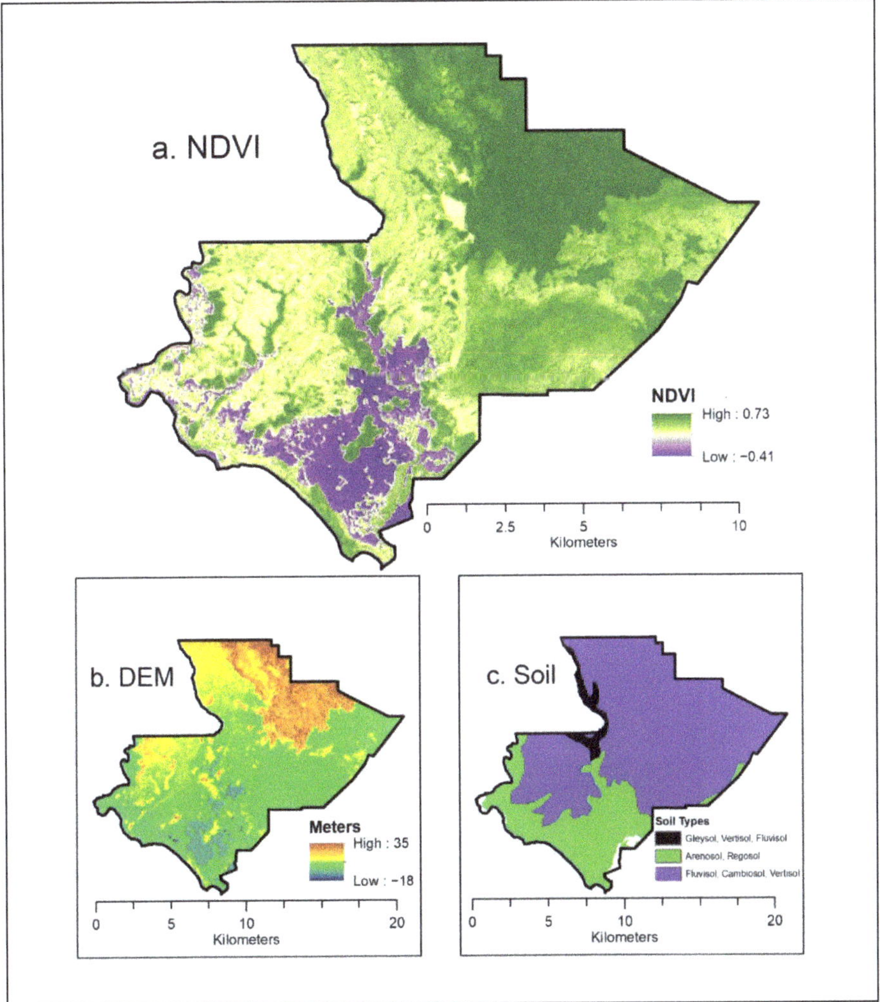

Figure 3. Maps of data used as variables in the Support Vector Machine (SVM) classification: (**a**) Normalized Difference Vegetation Index (NDVI) map of Landsat-5 Thematic Mapper (TM) image from March 2011; (**b**) Shuttle Radar Topography Mission (SRTM) Digital Elevation Model (DEM) showing elevation in meters of PCNP, from United States Geological Survey (USGS) Earth Explorer; (**c**) soil type map showing areas in PCNP with differing major soil types, from Selva Maya consortium.

2.4. Land Cover Classification and Change

2.4.1. Land Cover Classes

Other research carried out in Belize, including the southern coastal plain, identify the dominant land covers of PCNP as open savanna, dense tree savanna, forest and wetland [28,33]. The open savanna classification is characterized by the domination of grasses and sedges with scattered trees and shrubs. Similarly, the species composition of the dense tree savanna is essentially the same as the open savanna, but with semi-closed oak and pine canopies. The forest land cover includes lowland broad-leaved wet forests, lowland broad-leaved moist scrub forests, and mangrove and littoral forests.

Wetland land cover in PCNP are comprised of tall herb lowland swamp and Eleocharis marshes [33,34]. The focus of this study, however, is to characterize conversion to and from woody and non-woody vegetation over time, so the land cover classes used in this classification are more general: forest, grass, wetland, and water.

2.4.2. Training Data

Since fieldwork was not undertaken, all training and testing data in this study were collected via easily accessible high-resolution satellite imagery. Using the "Generate Random Points" tool in ArcMap, random points were generated throughout the PCNP shapefile. Using imagery from Google Earth and the ArcGIS satellite basemap, the random points were identified as belonging to one of the four training sample categories: forest, grass, wetland, or water. The 2017 update of the Belize Ecosystems map from Meerman and Sabido [38] and the Savanna Ecosystems Map of Belize from Bridgewater et al. [33] were used to aid in the spectral differentiation of classes. Additional classes for clouds and cloud shadows were used in the images from 1993 and 2011. A total of 198 points comprised the training dataset, with 133 of those being for the four land cover classes and 65 for clouds and shadows. Approximately 80% of the samples were used to train the classifier and 20% were saved for accuracy assessment [12]. The training and testing data were randomly split by the "Subset Features" tool in ArcMap.

2.4.3. Support Vector Machine Classification

A radial basis function kernel SVM was chosen for these classification maps due to the capability of the classifier to distinguish non-linear boundaries between classes, especially when training data sample size is low. The SVM classifier is a supervised classification technique that uses statistical learning to discriminate classes by maximizing the separation between discrete classes based on training data. The optimal separating hyperplane refers to this boundary, which maximizes class separation and minimizes misclassification [43,44].

The SVM classification was carried out in ENVI 5.5 using a radial basis function (RBF) kernel. The four classifications were carried out on dry season images using Landsat-2 MSS bands 1–4 for March 1975, Landsat-4 TM bands 1–7 for March 1993, Landsat-5 TM band 1–7 for March 2011, and Landsat-8 OLI/TIRS bands 1–8, 10 and 11 for January 2019 imagery. All four classifications also included the derived NDVI using the NIR and Red bands, the SRTM DEM, and the Belize Soils Map. The raw Landsat bands, NDVI, DEM and Belize Soils Map were all subset to the PCNP shapefile in ArcMap and stacked together as a multi-band file in ENVI. Accuracy assessments were performed in ENVI for all four classification maps using ground truth regions of interest.

2.4.4. Change Detection

The first step for detecting conversions to and from woody and non-woody vegetation was to simplify the land cover classes. Using the "Reclassify" tool in ArcMap 10.7.1, the forest class pixels were reclassified as "Woody" while grass, wetland, and water were reclassified as "Non-Woody." The cloud and cloud shadow pixels were reclassified to no data. After all maps were reclassified, we used the ArcMap "Raster Calculator" to create change trajectories. A change trajectory was created for all four dates to characterize overall woody and non-woody conversions in the time-series. Three additional change trajectories were created to show woody and non-woody conversion between 1975 and 1993, 1993 and 2011, and 2011 and 2019.

3. Results

3.1. Land Cover Classification and Change Trajectories

3.1.1. Classification Maps

The first output from the SVM classification is a land cover classification map for each of the four dates (Figure 4). These maps show where forest, grass, water, and wetland classes are located in PCNP. The 1993 and 2011 maps also contain cloud and cloud shadow classes. Figure 4 shows that, in all four maps, the four land cover classes are generally grouped together, but with transitional areas between them. Higher elevations and improved drainage promote closed forest in the northeast; seasonal inundations and nutrient-poor soils promote grasses and interspersed trees in the central and western low-lying areas; waterlogged soils promote wetlands in the east that border the grass, closed forest, and Atlantic Ocean to the south; water accumulates and discharges into the Caribbean Sea in the southwest and southeast. The most notable transitions are the more gradual transitions between grass and wetland, forest and wetland, and the quicker transition between grass and forest.

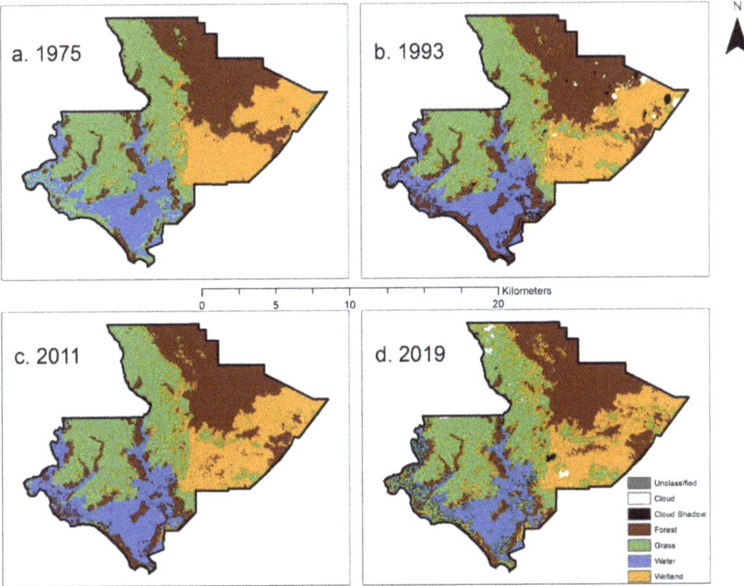

Figure 4. Support Vector Machine classification results based on Landsat imagery and other data for: (**a**) 1975; (**b**) 1993; (**c**) 2011; (**d**) 2019.

The error matrices (Table 1) show where misclassification occurred in the four images. Table 1 also shows class error commission and omission percentages, and overall accuracy and kappa coefficients. The overall accuracy for the 1975, 1993, 2011 and 2019 classifications are 88.5%, 96.9%, 96.2% and 86.1%, respectively, and the kappa statistics are 0.8402, 0.9612, 0.9474 and 0.8289. While the classifications perform well overall, some confusion, especially in the grass and wetland classes, can be noted visually. The transitional areas between grass and wetland are most well defined in the 1975 land cover classification, with each subsequent map having less definition between classes. Another issue for the SVM classifier was differentiating between spectrally similar water and cloud shadow, especially in the 2019 land cover classification; nevertheless, grouping classes together through reclassification results in new classification maps where these issues are less pronounced.

Table 1. Error matrices for Support Vector Machine classification results based on Landsat imagery and other data for 1975, 1993, 2011 and 2019. The 1975 and 2011 images were completely cloud-free.

Error Matrix	Forest	Grass	Water	Wetland	Cloud	Cloud Shadow	Total	Class Error Commission %	Class Error Omission %
					1975				
Unclassified	0	0	0	0	n/a	n/a	0		
Forest	6	0	0	0	n/a	n/a	6	0.0	0.0
Grass	0	7	2	0	n/a	n/a	9	22.2	12.5
Water	0	0	2	0	n/a	n/a	2	0.0	50.0
Wetland	0	1	0	8	n/a	n/a	9	11.1	0.0
Total	6	8	4	8	n/a	n/a	26		
					1993				
Unclassified	0	0	0	0	0	0	0		
Forest	6	0	1	0	0	0	7	14.3	0.0
Grass	0	8	0	0	0	0	8	0.0	0.0
Water	0	0	3	0	0	0	3	0.0	25.0
Wetland	0	0	0	8	0	0	8	0.0	0.0
Cloud	0	0	0	0	3	0	3	0.0	0.0
Cloud Shadow	0	0	0	0	0	3	3	0.0	0.0
Total	6	8	4	8	3	3	32		
					2011				
Unclassified	0	0	0	0	n/a	n/a	0		
Forest	6	0	1	0	n/a	n/a	7	14.3	0.0
Grass	0	8	0	0	n/a	n/a	8	0.0	0.0
Water	0	0	3	0	n/a	n/a	3	0.0	25.0
Wetland	0	0	0	8	n/a	n/a	8	0.0	0.0
Total	6	8	4	8	n/a	n/a	26		
					2019				
Unclassified	0	0	0	0	0	0	0		
Forest	6	0	0	0	0	0	6	0.0	0.0
Grass	0	7	2	0	0	0	9	22.2	12.5
Water	0	0	2	0	0	0	2	0.0	50.0
Wetland	0	1	0	8	2	0	11	27.3	0.0
Cloud	0	0	0	0	4	0	4	0.0	33.3
Cloud Shadow	0	0	0	0	0	4	4	0.0	0.0
Total	6	8	4	8	6	4	36		

1975 Kappa Coefficient: 0.8402 Overall Accuracy: 88.5%; 1993 Kappa Coefficient: 0.9612 Overall Accuracy: 96.9%;
2011 Kappa Coefficient: 0.9474 Overall Accuracy: 96.2%; 2019 Kappa Coefficient: 0.8289 Overall Accuracy: 86.1%

3.1.2. Change Trajectories

Figure 5 shows the four SVM classification maps after being reclassified. There is an average of 176,765 reclassified pixels in the four maps with approximately 70% being non-woody and 30% woody. The number of pixels in the four classifications are not equal because of the presence of cloud and cloud shadow pixels in the 1993 and 2011 imagery, which were reclassified as no data. Figure 6 shows the overall change trajectory using all SVM classifications in the time-series. Visually, most pixels did not undergo change, as the WWWW (brown) and NNNN (blue) trajectories represent pixels that, throughout the time-series, remain woody or non-woody, respectively. Table 2 shows that 86.7% of pixels did not undergo conversion throughout the entire time-series, while all other pixels underwent conversion at least once. Of the 13.3% that underwent conversion, 66.3% ended the time-series as non-woody and 33.7% as woody. There are some visually discernible spatial patterns to the conversions that occurred in the overall change trajectory, indicating that some portions of PCNP may be more susceptible to land cover change. Figure 6 shows that changes occurred along the coast in the south and southwest. Conversions also appear to be more frequent on the fringes of the pockets of forest that surround fluvial systems. Substantial change also occurred in the eastern PCNP wetland and within the northern eastern closed forest. It appears that little widespread change occurred in the grasslands, but with pronounced conversions occurring in transition zones between woody and non-woody vegetation.

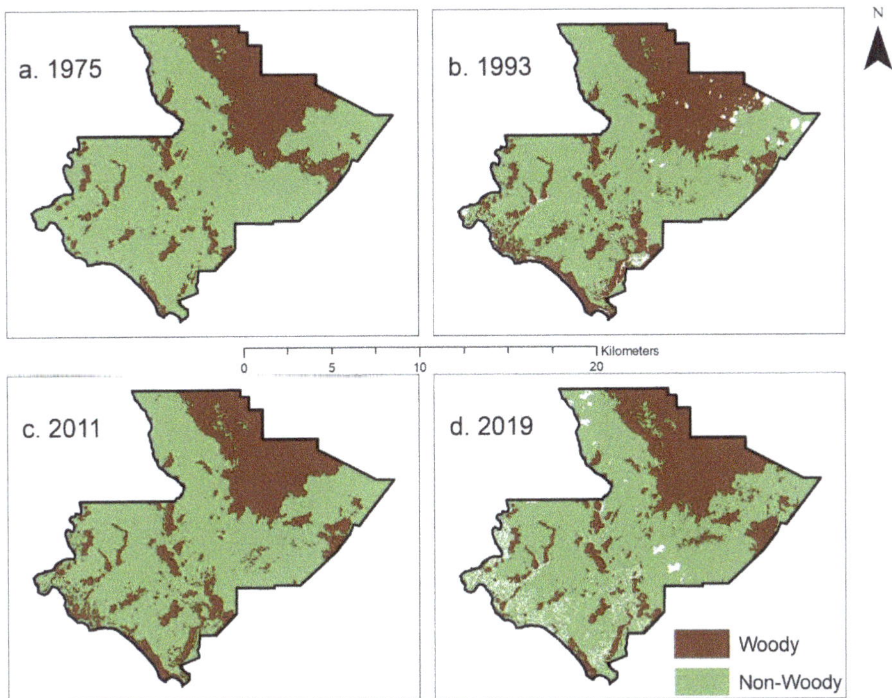

Figure 5. Maps showing the SVM land cover classification after reclassifying pixels as "Woody" or "Non-Woody": (**a**) 1975; (**b**) 1993; (**c**) 2011; (**d**) 2019. White pixels represent cloud and cloud shadow, which were reclassified as no data.

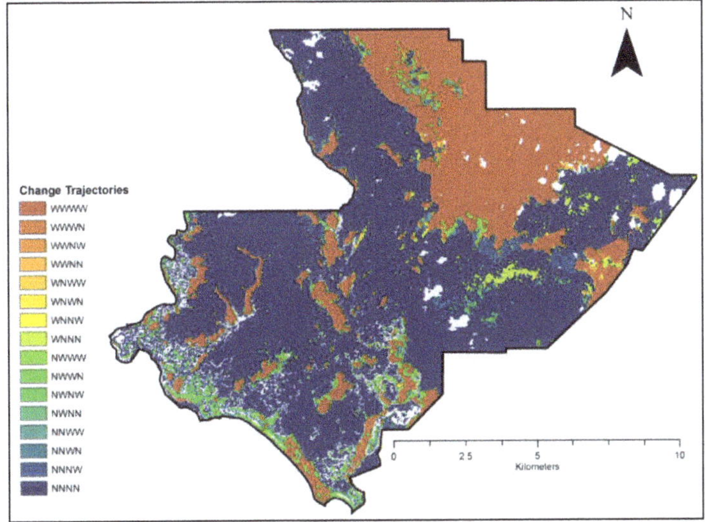

Figure 6. Overall change trajectory map using all four (1975, 1993, 2011, 2019) land cover classifications. Woody is "W" and non-woody is "N." White pixels represent no data.

Table 2. SVM classifier change trajectory pixel counts and percent change with explanation of each change trajectory. Woody is "W" and non-woody is "N".

Trajectory	1975	1993	2011	2019	Pixel Count	% Change
WWWW	W	W	W	W	41296	24.4
WWWN	N	W	W	W	3546	2.1
WWNW	W	N	W	W	398	0.2
WWNN	N	N	W	W	423	0.2
WNWW	W	W	N	W	398	0.2
WNWN	N	W	N	W	418	0.2
WNNW	W	N	N	W	258	0.2
WNNN	N	N	N	W	2134	1.3
NWWW	W	W	W	N	2060	1.2
NWWN	N	W	W	N	4231	2.5
NWNW	W	N	W	N	181	0.1
NWNN	N	N	W	N	1181	0.7
NNWW	W	W	N	N	1172	0.7
NNWN	N	W	N	N	3091	1.8
NNNW	W	N	N	N	2988	1.8
NNNN	N	N	N	N	105445	62.3
Total					169220	100.0

While the overall change trajectory map shows a more holistic view of change in the time-series, showing only the conversions between successive imagery dates allows for highlighting the timing of certain changes. Figure 7 shows three change trajectories which highlight woody and non-woody conversions between adjacent imagery dates. These maps make the timing of certain conversions much clearer. The conversions between 1975 and 1993 are perhaps the most distinct, with the south and southwest portions of PCNP undergoing widespread woody conversion (colored in yellow) along the coast. Non-woody conversion (green) occurred in the east in the transition from forest to wetland, indicating that these pixels may have converted from forest to wetland. Table 3 shows that 6.6% of all pixels converted to woody vegetation between 1975 and 1993, which is the highest rate of woody conversion between all four imagery dates. The 2.2% conversion to non-woody land cover between 1975 and 1993 is the lowest of the three change trajectories. Between 1993 and 2011, more than 95% of all pixels underwent no change and 3% converted to non-woody. Spatially, most of this change occurred along the coast, but there are also non-woody conversions within the grassland, wetland, and closed forest, especially in transitional areas. The conversions that occurred between 2011 and 2019 were mostly non-woody and appear to be in areas that had already undergone conversion at other points in the time-series.

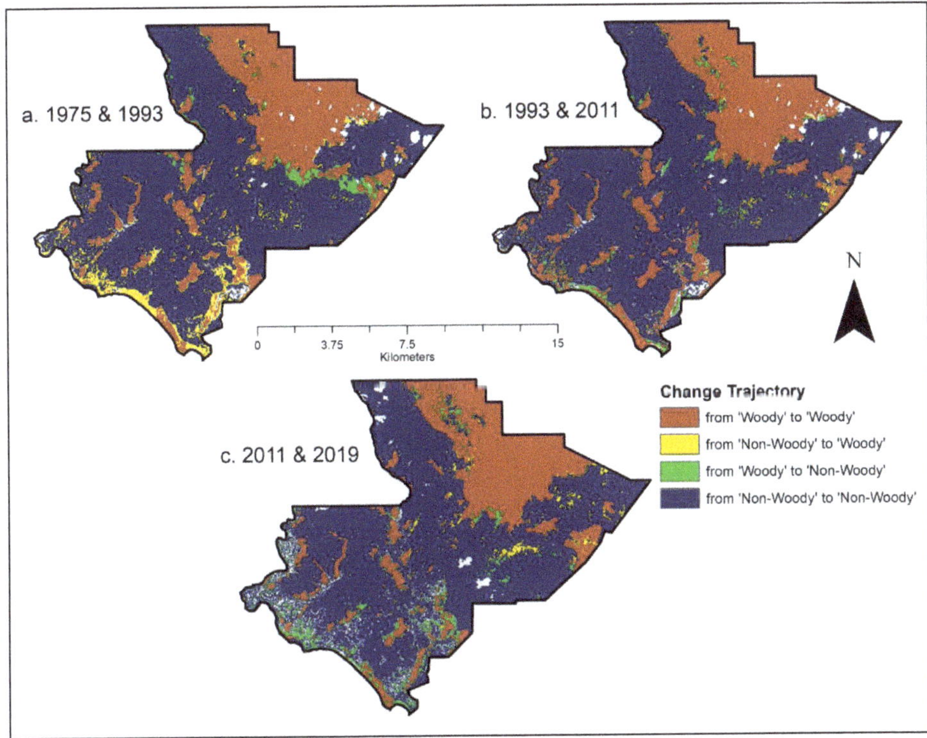

Figure 7. Change trajectory maps for successive imagery dates: (**a**) 1975 and 1993; (**b**) 1993 and 2011; (**c**) 2011 and 2019. White pixels represent no data.

Table 3. Change trajectory percent changes between successive imagery dates.

Trajectory	SVM Classifier % Change		
	1975–1993	**1993–2011**	**2011–2019**
Non-Woody to Woody	6.6	1.3	1.9
No Change	91.2	95.7	93.3
Woody to Non-Woody	2.2	3.0	4.8
Total	100.0	100.0	100.0

4. Discussion

This study demonstrates a novel application of a non-parametric supervised classification technique in a heterogeneous neotropical savanna landscape. SVM classifiers are especially useful when training data is limited and boundaries between classes are not abrupt, which are both true for this study. The SVM classification of the heterogenous landscape in PCNP is high-performing and performs similarly to other studies that use SVM to classify savanna vegetation [19,45,46]. This research adds to the growing recognition of the robustness of SVM classifiers in savannas and other heterogenous landscapes. The use of SVM needs to be explored further in savanna landscapes, especially in other small Central American savannas that are underrepresented in regional and global studies. Understanding conversions to and from woody and non-woody vegetation is a simple, but important way for researchers and land managers to compartmentalize and interpret change occurring on a landscape. This streamlined view of vegetation change is especially useful in savanna landscapes, which contain grass and trees in alternative stable states that are constantly experiencing change driven by disturbances and feedbacks [6]. Understanding where land conversions are occurring in savanna and adjacent landscapes is crucial so that managers can investigate the source of change and determine if planning and intervention is needed or if the change is part of the overall function of the ecosystem.

Our analysis reveals a relatively stable landscape in PCNP, but with spatially concentrated areas that are prone to vegetation conversion. The overall change trajectory (Figure 6, Table 2) does not reveal evidence of widespread woody or non-woody conversion in PCNP from 1975 to 2019. This finding is notable because it differs from studies of savanna vegetation dynamics on other continents. In sub-Saharan Africa, remote sensing studies have found evidence of woody encroachment in savannas [12,20]. The authors of [47] reviewed 16 studies that present evidence of woody encroachment and supplement those findings with continental-scale remote sensing analysis. Moreover, at a continental scale, a connection between mean annual precipitation and woody encroachment was discovered by the authors of [48], who also suggested that future increases of water availability in African Savannas could lead to enhanced woody conversion. In a protected area in northern Australia, researchers found forest expansion and grassland contraction in a lowland Eucalyptus savanna [49]. Further evidence of tropical rain forest expansion into savanna was found across multiple sites in northeast Australia, most likely attributed to increased atmospheric CO_2 [50]. In South America, it has been suggested that the expansion of riparian forests within the Brazilian Cerrado savannas has been occurring for thousands of years and will continue to do so with modern climate change [51]. Based on an extensive review covering 112 savanna sites, the authors of [5] conclude that widespread woody encroachment is occurring in savannas across Africa, Australia, and South America; however, regionally specific characteristics of plant biology, land use, disturbance, moisture availability, soil nutrients, etc. determine encroachment susceptibility for the landscape.

Figure 7 shows where conversions occurred between imagery dates and reveals that certain areas in PCNP are especially sensitive to change and have changed multiple times. These sensitive areas are along the coast and the transition zones between forest, grassland, and wetland. We hypothesize that some of these areas are especially susceptible to conversion due to disturbance and the inherent sensitivity and heterogeneity of vegetation transition zones in savannas. Clusters of conversion to non-woody vegetation (colored in green) between 1975 and 1993 may be a result of selective logging of broadleaf and pine trees, which occurred until becoming a protected area in 1994 [52]. Later in the time-series, non-woody conversions surrounding pockets of forest could be a result of fire management practices. TIDE uses prescribed fires in PCNP to maintain the pine savanna and promote post-fire pine regeneration. If the fire management efforts were not effective, it is likely that the 1993–2011 and 2011–2019 change trajectories would show an expansion of non-woody vegetation or woody encroachment in the grassland [52].

Between 1975 and 1993, there is also an area of concentrated woody conversion (colored in yellow) that occurs along the PCNP coastline. Meerman and Sabido [2001] indicate that this coastal area contains mangrove and littoral forest. Mangrove and littoral forest are dynamic environments that occupy the space between marine and terrestrial ecosystems, so it is not surprising that the PCNP coastline has experienced change in the time-series. Nevertheless, the spatially concentrated woody conversion along the coast suggests that a disturbance event, such as a hurricane or wildfire, or its aftermath could be the source of change. Since this a forest in the littoral zone, pre-1975 hurricane activity is more likely to have damaged the mangroves or reduced stand density. There were multiple tropical cyclone events that affected Belize before 1975: Hurricane Hattie (1961), Hurricane Francelia (1969), Hurricane Edith (1971), Tropical Storm Laura (1971), and Hurricane Fifi (1974) [53]. Stoddart [54] observed mangrove defoliation and widespread destruction of coconut palms in areas affected by Hurricane Hattie [54]. These pre-1975 events may have reduced littoral forest density and induced regrowth. The fact that there are also groups of pixels along the coast that went from woody to woody between 1975 and 1993 suggests that woody conversion was already happening before 1975 and continued through 1993. Even though mangrove forests are not typically associated with savanna landscapes, the importance of the mangroves to the pine savanna manifests through the interception of energy from hurricanes and other storms [55].

One of the contributions of this study to savanna science is the importance of ascertaining and including fine-scale driving factors, such as topography and soil type, into remote sensing analyses. Each savanna landscape is unique and contains varying drivers and disturbances that determine vegetation distribution, with common drivers often being plant available moisture and nutrients [56]. In the lowland savannas of Belize, edaphic and topographic factors play key roles in vegetation distribution [23]. These factors vary on multiple scales in Belize, and the fine-scale variations within these are important vegetation drivers in PCNP. Elevation ranges from 35 m to sea level and broadly distributes vegetation in PCNP; nevertheless, it appears that the fine-scale consequences of elevation, such as locally improved drainage, determine where woody vegetation can survive. Since there are widespread nutrient-poor soils in the coastal plain, pockets of nutrient rich, alluvial soils along fluvial systems also determine the extent to which woody vegetation can establish. As is seen in the SVM classification maps, pixels at higher elevations and around fluvial systems were classified as woody vegetation, which agrees with other studies [28,33,38]. These results parallel the findings of a study in Kenya by Coughenour and Ellis [57]. They determined that water availability and physical landscape pattern control the distribution of grass and trees, as woody vegetation almost only occurred in riparian zones. It was concluded that the there is a hierarchy of physical factors that determine where woody vegetation can develop on this landscape at multiple scales: climate at regional and continental, topographic influence on water redistribution and the geomorphic effect on soil moisture regionally, and water redistribution and disturbance at the local scale [57]. Similarly, the interaction between rainfall and topography distribute vegetation in the Serengeti-Maasai Mara savanna ecosystem [58]. Including the SRTM DEM and the soils map as variables in the SVM classifications were paramount in encompassing "bottom-up" drivers of woody and non-woody vegetation distribution that remotely sensed imagery alone cannot include. Future vegetation studies in PCNP should quantitatively describe the influence of riparian zones and alluvium on woody vegetation distribution.

5. Conclusions

The research presented here demonstrates that a non-parametric supervised land cover classification technique performs well in this heterogenous savanna landscape. Using a SVM classification and time-series change trajectories, we characterized woody and non-woody conversion in PCNP from 1975 to 2019 and found a relatively stable landscape. Our land cover classification revealed that some portions of PCNP that are susceptible to change tend to be in areas where the dominant vegetation cover is transitioning. These areas overlap with changes in elevation and/or riparian zones, indicating the importance of slight variations in drainage and nutrient availability. Consequently, our analysis suggests that topo-edaphic factors play a key role in controlling woody biomass in PCNP, and future studies should explore this further. We attribute the other areas with similarly grouped conversions to be disturbance or human induced, such as fire, hurricane-induced vegetation damage, or logging. Unlike many other savanna landscapes, woody encroachment is not widespread in PCNP.

The findings from this study have important implications for savanna scientists and land managers in Belize, such as a better understanding of fine-scale savanna vegetation drivers and areas susceptible to vegetation conversion. Adding to the knowledge of how this landscape functions will allow land managers to improve their efforts of protecting the pine savanna and its associated wildlife. Future work should improve the vegetation classification with field validation, quantitatively relate topo-edaphic factors to vegetation conversion, and compare these findings to other Central American savannas.

Author Contributions: Conceptualization, L.B. and H.V.H; methodology, L.B. and H.V.H; software, L.B.; formal analysis, L.B.; investigation, L.B.; resources, L.B.; data curation, L.B. and H.V.H; writing—original draft preparation, L.B.; writing—review and editing, L.B. and H.V.H; visualization, L.B.; supervision, H.V.H.; project administration, H.V.H. All authors have read and agreed to the published version of the manuscript.

Funding: This research received no external funding.

Acknowledgments: We would like to thank Cathy Smith for introducing us to this landscape and Michael Dobbins for providing insight about the study area.

Conflicts of Interest: The authors declare no conflict of interest.

References

1. Kottek, M.; Grieser, J.; Beck, C.; Rudolf, B.; Rubel, F. World Map of the Köppen-Geiger climate classification updated. *METZ* **2006**, *15*, 259–263. [CrossRef]
2. Chapin, F.S.; Matson, P.A.; Vitousek, P. *Principles of Terrestrial Ecosystem Ecology*; Springer Science & Business Media: Berlin/Heidelberg, Germany, 2011; ISBN 978-1-4419-9504-9.
3. Lehmann, C.E.R.; Anderson, T.M.; Sankaran, M.; Higgins, S.I.; Archibald, S.; Hoffmann, W.A.; Hanan, N.P.; Williams, R.J.; Fensham, R.J.; Felfili, J.; et al. Savanna Vegetation-Fire-Climate Relationships Differ Among Continents. *Science* **2014**, *343*, 548–552. [CrossRef]
4. Cole, M.M. The savannas. *Prog. Phys. Geogr. Earth Environ.* **1987**, *11*, 334–355. [CrossRef]
5. Stevens, N.; Lehmann, C.E.R.; Murphy, B.P.; Durigan, G. Savanna woody encroachment is widespread across three continents. *Glob. Chang. Biol.* **2017**, *23*, 235–244. [CrossRef]
6. Murphy, B.P.; Bowman, D.M.J.S. What controls the distribution of tropical forest and savanna? *Ecol. Lett.* **2012**, *15*, 748–758. [CrossRef] [PubMed]
7. Staver, A.C.; Archibald, S.; Levin, S.A. The Global Extent and Determinants of Savanna and Forest as Alternative Biome States. *Science* **2011**, *334*, 230–232. [CrossRef] [PubMed]
8. Southworth, J.; Zhu, L.; Bunting, E.; Ryan, S.J.; Herrero, H.; Waylen, P.R.; Hill, M.J. Changes in vegetation persistence across global savanna landscapes, 1982–2010. *J. Land Use Sci.* **2016**, *11*, 7–32. [CrossRef]
9. Tropical and subtropical grasslands, savannas and shrublands|Biomes|WWF. Available online: https://www.worldwildlife.org/biomes/tropical-and-subtropical-grasslands-savannas-and-shrublands (accessed on 1 May 2020).
10. Ferreira, J.N.; Bustamante, M.M.d.C.; Davidson, E.A. Linking woody species diversity with plant available water at a landscape scale in a Brazilian savanna. *J. Veg. Sci.* **2009**, *20*, 826–835. [CrossRef]
11. Dantas, V.d.L.; Batalha, M.A. Vegetation structure: Fine scale relationships with soil in a cerrado site. *Flora - Morphol. Distrib. Funct. Ecol. Plants* **2011**, *206*, 341–346. [CrossRef]
12. Herrero, H.V.; Southworth, J.; Bunting, E. Utilizing Multiple Lines of Evidence to Determine Landscape Degradation within Protected Area Landscapes: A Case Study of Chobe National Park, Botswana from 1982 to 2011. *Remote Sens.* **2016**, *8*, 623. [CrossRef]
13. Pennington, R.T.; Lehmann, C.E.R.; Rowland, L.M. Tropical savannas and dry forests. *Curr. Biol.* **2018**, *28*, R541–R545. [CrossRef] [PubMed]
14. Southworth, J.; Munroe, D.; Nagendra, H. Land cover change and landscape fragmentation—Comparing the utility of continuous and discrete analyses for a western Honduras region. *Agric. Ecosyst. Environ.* **2004**, *101*, 185–205. [CrossRef]
15. Jensen, J.R. *Introductory Digital Image Processing: A Remote Sensing Perspective*, 4th ed.; Prentice Hall Press: Upper Saddle River, NJ, USA, 2015; ISBN 978-0-13-405816-0.
16. Maximum Likelihood. Available online: https://www.harrisgeospatial.com/docs/MaximumLikelihood.html (accessed on 21 May 2020).
17. Support Vector Machine. Available online: https://www.harrisgeospatial.com/docs/SupportVectorMachine.html (accessed on 15 June 2020).
18. Decision Tree. Available online: https://www.harrisgeospatial.com/docs/DecisionTree.html (accessed on 15 June 2020).
19. Herrero, H.V.; Southworth, J.; Bunting, E.; Kohlhaas, R.R.; Child, B. Integrating Surface-Based Temperature and Vegetation Abundance Estimates into Land Cover Classifications for Conservation Efforts in Savanna Landscapes. *Sensors* **2019**, *19*, 3456. [CrossRef]
20. Vogel, M.; Strohbach, M. Monitoring of savanna degradation in Namibia using Landsat TM/ETM+ data. In Proceedings of the 2009 IEEE International Geoscience and Remote Sensing Symposium, Cape Town, South Africa, 12–17 July 2009; Volume 3, pp. III-931–III-934.

21. Furley, P.A. The nature and diversity of neotropical savanna vegetation with particular reference to the Brazilian cerrados. *Glob. Ecol. Biogeogr.* **1999**, *8*, 223–241. [CrossRef]
22. Mistry, J. Savannas. *Prog. Phys. Geogr. Earth Environ.* **2000**, *24*, 601–608. [CrossRef]
23. Furley, P.A. *Savannas: A Very Short Introduction*; Oxford University Press: Oxford, UK, 2016; ISBN 978-0-19-871722-5.
24. Biomes|Conserving Biomes|WWF. Available online: https://www.worldwildlife.org/biomes (accessed on 1 May 2020).
25. Garbulsky, M.F.; Paruelo, J.M. Remote sensing of protected areas to derive baseline vegetation functioning characteristics. *J. Veg. Sci.* **2004**, *15*, 711–720. [CrossRef]
26. Josefsson, T.; Hörnberg, G.; Östlund, L. Long-Term Human Impact and Vegetation Changes in a Boreal Forest Reserve: Implications for the Use of Protected Areas as Ecological References. *Ecosystems* **2009**, *12*, 1017–1036. [CrossRef]
27. April Sahara, E.; Sarr, D.A.; Van Kirk, R.W.; Jules, E.S. Quantifying habitat loss: Assessing tree encroachment into a serpentine savanna using dendroecology and remote sensing. *For. Ecol. Manag.* **2015**, *340*, 9–21. [CrossRef]
28. Stuart, N.; Barratt, T.; Place, C. Classifying the Neotropical savannas of Belize using remote sensing and ground survey. *J. Biogeogr.* **2006**, *33*, 476–490. [CrossRef]
29. Climate Summary—National Meteorological Service of Belize. Available online: https://www.hydromet.gov.bz/climatology/climate-summary (accessed on 18 May 2020).
30. Fick, S.E.; Hijmans, R.J. WorldClim 2: New 1-km spatial resolution climate surfaces for global land areas. *Int. J. Climatol.* **2017**, *37*, 4302–4315. [CrossRef]
31. Harris, I.; Jones, P.D.; Osborn, T.J.; Lister, D.H. Updated high-resolution grids of monthly climatic observations – the CRU TS3.10 Dataset. *Int. J. Climatol.* **2014**, *34*, 623–642. [CrossRef]
32. Funk, C.; Peterson, P.; Landsfeld, M.; Pedreros, D.; Verdin, J.; Shukla, S.; Husak, G.; Rowland, J.; Harrison, L.; Hoell, A.; et al. The climate hazards infrared precipitation with stations—A new environmental record for monitoring extremes. *Sci. Data* **2015**, *2*, 1–21. [CrossRef] [PubMed]
33. Cameron, I.; Stuart, N.; Goodwin, Z. *Savanna Ecosystems Map of Belize 2011: Technical Report*; Darwin Initiative Project 17022; University of Edinburgh: Edinburgh, Scotland, 2011; pp. 5–16.
34. Goodwin, Z.A.; Lopez, G.N.; Stuart, N.; Bridgewater, S.G.M.; Haston, E.M.; Cameron, I.D.; Michelakis, D.; Ratter, J.A.; Furley, P.A.; Kay, E.; et al. A checklist of the vascular plants of the lowland savannas of Belize, Central America. *Phytotaxa* **2013**, *101*, 1–119. [CrossRef]
35. Donoghue, S.; Furley, P.A.; Stuart, N.; Haggis, R.; Trevaskis, A.; Lopez, G. The nature and spatial variability of lowland savanna soils: Improving the resolution of soil properties to support land management policy. *Soil Use Manag.* **2019**, *35*, 547–560. [CrossRef]
36. Myers, R.; O'Brien, J.; Morrison, S. *Fire Management Overview of the Caribbean Pine (Pinus caribaea) Savannas of the Mosquitia, Honduras*; The Nature Conservancy: Arlington, VA, USA, 2006; p. 6.
37. Fire Management. Available online: http://tidebelize.org/fire-management/ (accessed on 17 May 2020).
38. Meerman, J.; Sabido, W. *Central American Ecosystems Map: Belize*; Programme for Belize: Belize City, Belize, 2001.
39. EarthExplorer. Available online: https://earthexplorer.usgs.gov/ (accessed on 17 June 2020).
40. Measuring Vegetation (NDVI & EVI). Available online: https://earthobservatory.nasa.gov/features/MeasuringVegetation (accessed on 18 May 2020).
41. Bai, Z.G.; Dent, D.L.; Olsson, L.; Schaepman, M.E. Proxy global assessment of land degradation. *Soil Use Manag.* **2008**, *24*, 223–234. [CrossRef]
42. Selva Maya Consortium Soils of Belize. Available online: https://databasin.org/datasets/21a4f58393904edcbdb1ae031a4c6b68 (accessed on 11 May 2020).
43. Cortes, C.; Vapnik, V. Support-vector networks. *Mach Learn* **1995**, *20*, 273–297. [CrossRef]
44. Burges, C.J.C. A Tutorial on Support Vector Machines for Pattern Recognition. *Data Min. Knowl. Discov.* **1998**, *2*, 121–167. [CrossRef]
45. Paneque-Gálvez, J.; Mas, J.-F.; Moré, G.; Cristóbal, J.; Orta-Martínez, M.; Luz, A.C.; Guèze, M.; Macía, M.J.; Reyes-García, V. Enhanced land use/cover classification of heterogeneous tropical landscapes using support vector machines and textural homogeneity. *Int. J. Appl. Earth Obs. Geoinf.* **2013**, *23*, 372–383. [CrossRef]
46. Silva, G.B.S.; Mello, M.P.; Shimabukuro, Y.E.; Rudorff, B.F.T.; de Castro Victoria, D. Multitemporal classification of natural vegetation cover in Brazilian Cerrado. In Proceedings of the 2011 6th International Workshop on the Analysis of Multi-temporal Remote Sensing Images (Multi-Temp), Trento, Italy, 12–14 July 2011; pp. 117–120.

47. Mitchard, E.T.A.; Flintrop, C.M. Woody encroachment and forest degradation in sub-Saharan Africa's woodlands and savannas 1982–2006. *Philos. Trans. R. Soc. B Biol. Sci.* **2013**, *368*, 20120406. [CrossRef]
48. Sankaran, M.; Hanan, N.P.; Scholes, R.J.; Ratnam, J.; Augustine, D.J.; Cade, B.S.; Gignoux, J.; Higgins, S.I.; Le Roux, X.; Ludwig, F.; et al. Determinants of woody cover in African savannas. *Nature* **2005**, *438*, 846–849. [CrossRef]
49. Bowman, D.M.J.S.; Walsh, A.; Milne, D.J. Forest expansion and grassland contraction within a Eucalyptus savanna matrix between 1941 and 1994 at Litchfield National Park in the Australian monsoon tropics. *Glob. Ecol. Biogeogr.* **2001**, *10*, 535–548. [CrossRef]
50. Tng, D.Y.P.; Murphy, B.P.; Weber, E.; Sanders, G.; Williamson, G.J.; Kemp, J.; Bowman, D.M.J.S. Humid tropical rain forest has expanded into eucalypt forest and savanna over the last 50 years. *Ecol. Evol.* **2012**, *2*, 34–45. [CrossRef] [PubMed]
51. Silva, L.C.R.; Sternberg, L.; Haridasan, M.; Hoffmann, W.A.; Miralles-Wilhelm, F.; Franco, A.C. Expansion of gallery forests into central Brazilian savannas. *Glob. Chang. Biol.* **2008**, *14*, 2108–2118. [CrossRef]
52. *Report For: TIDE Payne's Creek National Park Biodiversity Assessment*; Wildtracks: Belize City, Belize, 2005; pp. 4–18.
53. Hurricanes and Tropical Storms Affecting Belize since 1930. Available online: http://consejo.bz/weather/storms.html (accessed on 18 May 2020).
54. Stoddart, D.R. Catastrophic Storm Effects on the British Honduras Reefs and Cays. *Nature* **1962**, *196*, 512–515. [CrossRef]
55. Murray, M.R.; Zisman, S.A.; Furley, P.A.; Munro, D.M.; Gibson, J.; Ratter, J.; Bridgewater, S.; Minty, C.D.; Place, C.J. The mangroves of Belize: Part 1. distribution, composition and classification. *For. Ecol. Manag.* **2003**, *174*, 265–279. [CrossRef]
56. Furley, P. Plant ecology, soil environments and dynamic change in tropical savannas. *Prog. Phys. Geogr. Earth Environ.* **1997**, *21*, 257–284. [CrossRef]
57. Coughenour, M.B.; Ellis, J.E. Landscape and Climatic Control of Woody Vegetation in a Dry Tropical Ecosystem: Turkana District, Kenya. *J. Biogeogr.* **1993**, *20*, 383–398. [CrossRef]
58. Reed, D.N.; Anderson, T.M.; Dempewolf, J.; Metzger, K.; Serneels, S. The spatial distribution of vegetation types in the Serengeti ecosystem: The influence of rainfall and topographic relief on vegetation patch characteristics. *J. Biogeogr.* **2009**, *36*, 770–782. [CrossRef]

Commentary

Socio-Environmental Dynamics of Alpine Grasslands, Steppes and Meadows of the Qinghai–Tibetan Plateau, China: A Commentary

Haiying Feng [1,*] and Victor R. Squires [2,*]

1 Qinzhou Development Research Institute, Beibu Gulf University, Qinzhou 535011, China
2 Faculty of Natural Resources, Formerly University of Adelaide, Adelaide, SA 5005, Australia
* Correspondence: fhy@bbgu.edu.cn (H.F.); squires200@yahoo.com.au (V.R.S.)

Received: 24 March 2020; Accepted: 15 May 2020; Published: 17 September 2020

Abstract: Alpine grasslands are a common feature on the extensive (2.6 million km^2) Qinghai–Tibet plateau in western and southwestern China. These grasslands are characterized by their ability to thrive at high altitudes and in areas with short growing seasons and low humidity. Alpine steppe and alpine meadow are the principal plant Formations supporting a rich species mix of grass and forb species, many of them endemic. Alpine grasslands are the mainstay of pastoralism where yaks and hardy Tibetan sheep and Bactrian camels are the favored livestock in the cold arid region. It is not only their importance to local semi nomadic herders, but their role as headwaters of nine major rivers that provide water to more than one billion people in China and in neighboring countries in south and south-east Asia and beyond. Grasslands in this region were heavily utilized in recent decades and are facing accelerated land degradation. Government and herder responses, although quite different, are being implemented as climate change and the transition to the market economy proceeds apace. Problems and prospects for alpine grasslands and the management regimes being imposed (including sedentarization, resettlement and global warming are briefly discussed.

Keywords: endemism; floristics; nomadic; yaks; livestock; culture; altitude; watersheds; rivers; biodiversity; grassland management; land degradation; rodents; policy; interventions

1. Introduction

This study is a brief overview and synthesis of key points relating to the dynamics of alpine grasslands on the high-altitude and generally arid Qinghai–Tibet Plateau (QTP). It presents a synthesis of what is known about the key issues relating to alpine grasslands and the people, both locally and in distant countries, who depend on the ecosystem goods and services the grasslands provide. Many people use the grassland daily for their lives and livelihoods in this remote and harsh environment. Its purpose is to introduce the region and highlight aspects of the biogeography, ecology, climatology and demography that make this region on the "Roof of the World" or the "Third Pole" such an important one that provides water to about two billion people. In China itself, hundreds of millions rely on the rivers that arise on the QTP to provide water for irrigation, industrial uses, navigation and domestic water supply. Many hundreds of millions of people who live in countries in South Asia and South-east Asia are utterly dependent on flows along the Mekong, Brahmaputra and the other major rivers that arise on the QTP. The QTP supports millions of semi-nomadic herders and their livestock in Qinghai and in Tibet where livestock such as yak, hardy Tibetan sheep and Bactrian camels provide the sole source of sustenance and income.

Regrettably, anthropogenic environmental degradation (Figure 1)—exacerbated by changing climate [1], is accelerating as populations of both people and livestock rise [2], and policy measures designed to protect biodiversity and watersheds are implemented. Climate change, overgrazing and other reasons caused continuous degradation of alpine grasslands (see below). Several categories of alpine grassland are recognized based on the botanical composition and species diversity (see below). The management regimes under which they are used has changed a lot in the past 80 years (see below) and land degradation is now a major concern.

Figure 1. Land degradation on the cold and arid Qinghai–Tibet Plateau has accelerated in response to rising pressure from increasing populations of people (inward migration, natural increase) and of livestock (photo: H.Y. Feng).

Context and Setting

The QTP is the largest upland in the world and encompasses Xizang (Tibet) Qinghai province and parts of northern Yunnan and Sichuan provinces of China (Figure 2). Much of QTP is at an elevation above 2500 m above sea level (m ASL) and it includes Mt Everest and other high peaks of the Himalayan chain of mountains where elevations exceed 7000 m ASL. Of course, grassland vegetation, except for subnival cushion plants such as *Thylacospermum caespitosum* that can grow at elevations up to 6000 m ASL, is absent at elevations of above about 4500 m ASL.

The total area is approximately 2.6 million km^2 (about the size of Argentina or twice the area of Germany). The QTP is surrounded by massive mountain ranges. The QTP is bordered to the south by the Himalayan Range, to the north by the Kunlun Range which separates the QTP from the Hexi Corridor and the Gobi Desert that occupies parts of western Gansu, Xinjiang and parts of western Inner Mongolia even extending into neighboring Mongolia [3].

To the east and southeast the plateau gives way to the mountainous headwaters of nine of Asia's most important rivers (Figure 3) including the Mekong, Yangtze, Yellow, Salween and Brahmaputra rivers (Figure 3b). The QTP is home to the world's highest mountains and some of the world's cold deserts. The QTP is a high altitude, arid steppe interspersed with mountain ranges and large brackish lakes. Despite the extreme climate there are extensive areas of grasslands. Grasslands occupy about 60% of the land surface area of the QTP. Alpine grassland is a major ecosystem on the Qinghai–Tibet Plateau [4] playing an important role as an eco-safety barrier and serving as the basis of highland animal husbandry (Figures 3a and 4).

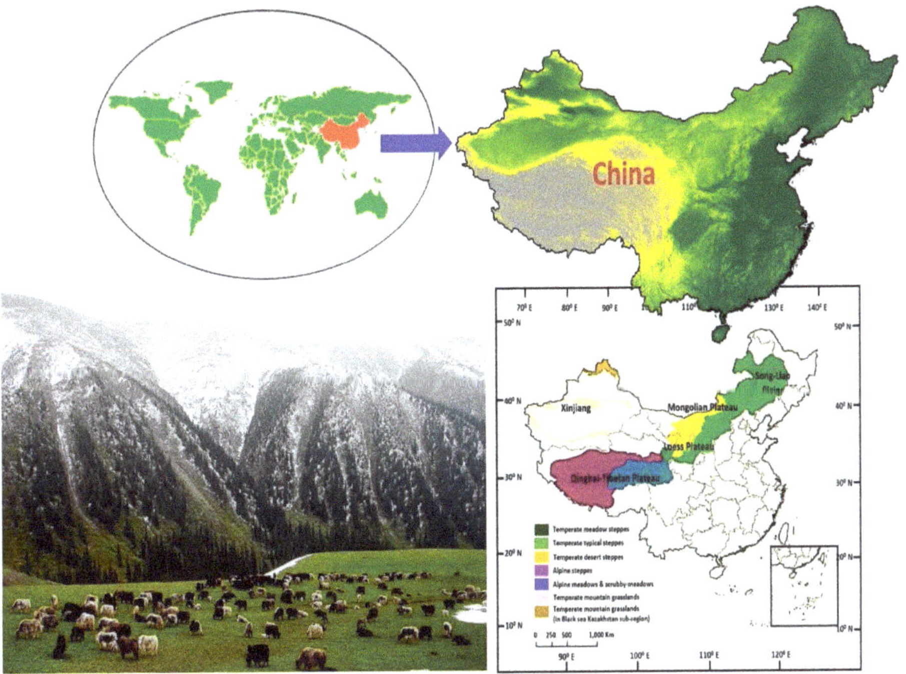

Figure 2. Maps showing right to left (clockwise) China (in red) in relation to the whole world, the location of the Qinghai–Tibet Plateau (QTP) in China (gray shaded area in SW of China) and the distribution inter alia of alpine steppes and alpine meadows on the QTP. The photo shows yaks grazing alpine grassland at an altitude of 4000 meters (photo: V.R. Squires).

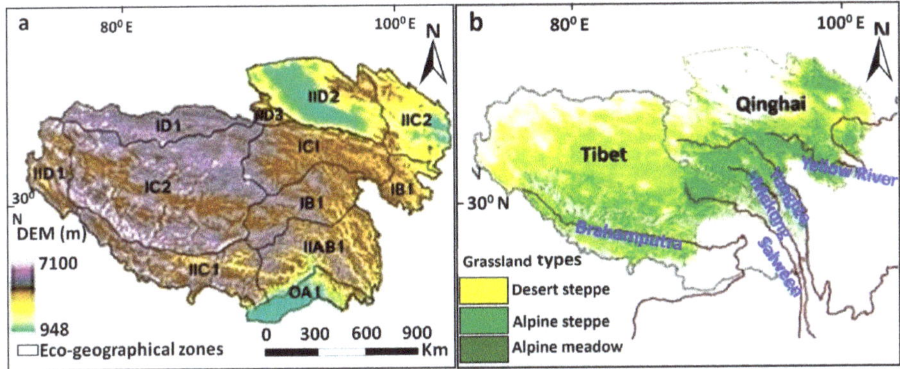

Figure 3. Map of Qinghai–Tibet Plateau showing (**a**) principal eco-geographical zones, elevation and (**b**) grassland types as well as the major rivers that arise on the Plateau. The relatively small areas of true alpine meadow are scattered throughout the lower altitude regions and are highly prized as grazing, especially by sheep herders (see Figure 4).

Figure 4. Alpine grasslands on the Qinghai–Tibet Plateau, especially alpine meadows, are prime grazing lands. Government policy is to regulate grazing pressure through imposition of grazing bans and restrictions on herder mobility (transhumance) (photo: V.R. Squires).

2. Origin of the Grasslands

In the recent review by Li et al. [5] it is argued that much of the contemporary natural grassland region was covered by subtropical forests and savanna woodlands during the Tertiary period. The grasslands formed as a consequence of the uplift of the QTP that blocked the monsoon rains from the south. The region became drier, favoring the gradual development of xerophilous steppes replacing the former mesophilous vegetation. The evolutionary history of biodiversity and endemism on the QTP is still debated, depending on differing hypotheses on highland uplift and climate history. Similarly, the history and intensity of human impact on Tibetan grasslands are still under discussion based on evidence from diverse disciplines [2,6,7].

The present day QTP supports various alpine vegetation types that include meadow, steppe, cold desert and subnival cushion plant communities [2,5] at elevations ranging from 3500 to nearly 6000 m above sea level (m ASL). The southern and eastern edges of the steppe have grasslands which can sustainably support populations of nomadic herdsmen. Frosts occur for six months of the year and there is permafrost under extensive parts of the region. Annual precipitation ranges from 100–300 mm and occurs mainly as hailstorms. Within the QTP, moisture determines whether a specific subregion supports meadow, steppe or alpine desert vegetation. Alpine (Here the word "alpine" does not strictly refer to a mountain climate which is not warm enough to allow for tree growth, here it indicates high-elevation distribution (see Box 1) Grassland is the dominant ecosystem covering over 50 per cent of the QTP [2,4,5,7]. Alpine steppes are commonly found on the QTP [8]. The grasslands on the QTP, including alpine steppes and meadows, are delimited as one alpine vegetation region [9,10]. The alpine steppes are distributed across the larger part of the QT Plateau, while the alpine meadows occur mainly on the southeastern parts, typically between 2500 and 4000 m ASL [10].

According to Li et al. [5] the QTP vegetation region has a very rich, diverse and often endemic flora. It comprises species of Central Asian (semi-) deserts, as well as temperate East Asian and Himalayan elements [8]. The dominant herbaceous species are not typically grasses but are frequently sedges from the genus *Kobresia* (Studies by Global Carex Group [11] suggest that the genus *Kobresia* should be included into *Carex*) and *Carex*, especially in alpine and subalpine meadows. A total of 5766 plant species have been documented for the Tibetan Autonomous Region [10]. More than 2000 of these species are recorded for the grassland region, of which 86% are forage plants and occur mainly in the more humid and sub-humid grasslands of eastern Tibet, while 540 species occur in the arid and semi-arid rangelands that lie further west [12,13]. The flora of alpine steppes consists of Eurasian steppe species (*Festuca kryloviana, F. pseudovina, Stipa glareosa, S. krylovii, S. bungaena*), as well as Central Asian alpine (*S. subsessiliflora*), Pamir–QTP (*S. purpurea, S. uroborowskyi, F. olgae, Carex moorcroftii*) and QTP endemics (*S. subsessiliflora* var. *basiplumosa* and *S. capillacea*) [11–13]. A 'snapshot' of the variety and distribution of key functional groups of alpine plants is shown in Table 1.

Table 1. Typical botanical composition in high elevation alpine grasslands # (percentage basis).

Functional Groups	Alpine Steppes			Alpine Desert Steppe		Alpine Meadow		Temperate Meadow Steppe
	Plains	Slopes	Mountains	Slopes	Mountains	Slopes	Mountains	Plains
Sedges/grasses	15.2/Tr.	28.5/Tr.	22.8/Tr.	48.5/Tr.	Tr./69.8	Tr.	40.70	Tr./3.87
Forbs (edible/inedible)	17.5	10.6	35.2	6.2	9.3	29.4	17.2	18.7/7.00 *
Legumes	Tr.	Tr.	Tr.	Tr.	4.05	2.45	Tr.	Tr.

Note. # data from sites scattered across Tibet (alpine steppe and alpine desert steppe) and Qinghai (alpine meadow and temperate meadow steppe). * Some forbs are toxic and/or unpalatable to livestock. In Temperate Meadow steppe 7.00 % is inedible. Tr. refers to trace amounts (<1%) Tr./3.87 refers to Tr. of sedge and 3.87 of grass. Data source: Feng, unpublished. This is a generalization based on numerous field surveys by H. Feng and her students over the past 10 years.

2.1. Understanding QTP Grassland Vegetation

From 1973 when the first comprehensive full-scale series of scientific expeditions for the QTP were initiated, much has been learned about the alpine grasslands of the QTP. Most data pertained to vegetation distribution using methods proposed by Ellenberg & Mueller-Dombois [14,15]. After 1992 several permanent monitoring stations were established at key sites to monitor eco-environmental changes on the QTP. A relatively complete monitoring and early warning system was set up that integrates air and land networks. A network developed for monitoring and studying earth surface processes and environment in extremely cold areas of high altitude. Observation networks exist that relate to environmental protection, land, agriculture, forestry, water conservancy, meteorology and some other fields. Under the Chinese ecosystem research network, observation stations (Figure 5) for eight different ecosystems, have been established on the QTP. Based on the earlier surveys and subsequent studies the following division of alpine grasslands has been suggested, as set out in 2.1 below. Meteorological stations and vegetation and land surface monitoring assists in devising revised policies for management of the vast areas of alpine pastures in the Study area (Qinghai–Tibet Plateau). Figure 3b shows the distribution of the three major alpine grassland types and there is a brief description of each, including listing of key plant species, in 3.1 below.

Our study area encompasses the vast tract of land (bigger than Spain) in southern and eastern QTP (Figure 5) although there were temporary field sites in the border areas near Tajikistan, Kyrgyzstan, Nepal, Bhutan and Myanmar. We did most of our botanical work on grasslands over a period of more than 12 years within a short radius of the monitoring stations (Figure 5) but our rural sociology work extended to herder communities further west into Tibet.

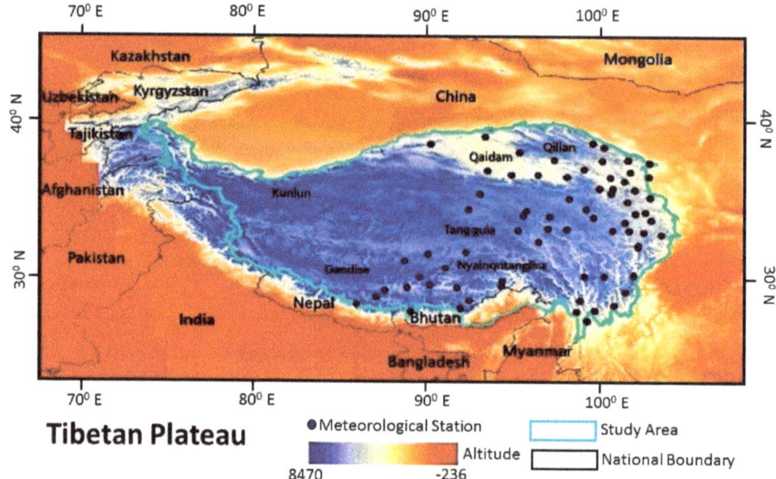

Figure 5. Qinghai–Tibet Plateau of China and the surrounding countries. Location of the monitoring stations is shown (Source: authors).

2.2. Types and Categories of Alpine Grasslands

Alpine steppes are mostly found at elevation between 3300 and 5300 m ASL on the QTP. The common dominant species are extremely cold- and drought-tolerant grasses and dwarf shrubs, with more or less pronounced admixture of cushion plants. The dominants include locally distributed species such as *Stipa subsessiliflora* var. *basiplumosa, S. roborowskyi, Festuca olgae, Carex moorcroftii, Orinus thoroldii O. kokonorica* and *Artemisia salsoloides* var. *wellbyi*; yet there also are common plants of Eurasian steppes such as *Festuca kryloviana* and *F. pseudovina* (Homophytic synonym is *Festuca pulchra*).

Marshy meadows are, according to [5] "primarily composed of hygrophilous herbs and have transitional characteristics between a typical meadow and a marsh. Their occurrence is generally associated with high soil moisture regulated by special topographic factors, most often at poorly drained sites, thus with humid and poorly aerated soil". The existence of a permafrost layer is also a major cause for association of humid soil conditions and marshy meadows, such as on the QTP. Massive layers of humus and even peat or a semi-peat layer accumulate in soils of marshy meadows due to slow decomposition of plant residues. Examples of alpine marshy meadows include those dominated by *Kobresia tibetica* along streams or waterlogged areas over permafrost in the catchments of the Yellow River, Yangtze, Mekong and Salween. *Kobresia schoenoides* is, however, dominant along the Yarlung Zangbo River. *Carex schmidtii* or *Sanguisorba parviflora* dominate at the river flats adjacent to forest stands in Hinggan mountains in the eastern QTP.

Saline meadows are mainly composed of mesic perennial halophytes, that is, salt tolerant or otherwise adapted species and occur on salinized low-lying sites, wide valleys, fringes of lake basins and river flats within steppe and desert regions. Most widely distributed saline meadows include those dominated by *Achnatherum splendens, Leymus dasystachys, Phragmites communis, Sophora alopecuroides, Crypsis aculeate* and *Suaeda* spp.

Alpine meadows occur mainly on the eastern QTP and in part on the high mountains of northwestern China and northern China, under a cold–arid climate with intermediate humidity, adequate sunshine, intense solar radiance and strong winds. As Box 1 shows, the Chinese interpretation of 'meadow' differs from the more common usage.

Box 1. Meadows as defined in the Chinese context.

Meadows in classic European understanding are agriculturally managed grasslands regularly mown for livestock forage (UNESCO classification) [14,15]. In China, meadow refers to natural grasslands primarily composed of mesophilous perennial herbs that thrive under moderate moisture conditions. The meadows naturally occupy vast continuous areas with moderate moisture conditions that derive from atmospheric precipitation and with relatively low temperature that limits the growth of trees, such as alpine meadows on the QTP. However, in most other cases, meadows are azonal, occurring in the alpine, montane, plain and coastal areas of the temperate zones, with moisture originating from groundwater, runoff or melted snow and ice; these meadows are classified into typical, marshy and salt meadows, based on the ecological traits of their constituent species, that is, the dominance of mesophilous, xero-mesophilous, hygro-mesophilous or halo-mesophilous species in their composition [10,16,17].

Alpine meadow soils are shaped by intense soil-forming processes during pedogenesis and are characterized by the occurrence of semi-decomposed residual turf, closely interwoven root mats and a poorly intermingled humus layer. The *Kobresia* alpine meadows are the most typical and most widespread of the alpine meadows in China [2,4,5]. *Kobresia* species are both mesophilous and cold-and drought-tolerant. There are about 40 *Kobresia* species, dominant ones include *K.pygmaea, K. humilis, K. capillifolia, K. graminifolia, K. setchwanensis, K. prattii, K. royleana, K. smirnovii* and *K. bellardii.* There are also alpine meadows dominated by sedges such as *Carex scabriostris, Blysmus sinocompressus,* bunchgrasses such as *Festuca ovina, F. pseudovina, Poa alpina, Anthoxanthum odoratum and* forbs such as *Polygonum viviparum, P. sphaerostachyum, Ligularia virgaurea* and *Alchemilla japonica.*

The grasslands have vertical zonality [17] from relatively low altitude (2400 m ASL) to high altitude (>4500 m ASL). There is also horizontal zonality from south-east QTP to the northwest region [4,13,17]. The sequence includes alpine meadow, alpine steppe, desert steppe and alpine desert that include sand dunes. The dominant plants in these grasslands include at the lower elevations *Poa alpine, Festuca rubra, F. ovina* and a suite of forbs in the genera *Anemone, Ranunculus, Stellaria.* On drier/higher sites *Kobresia humilis, Leymus chinensis, Stipa baicalensis, Stipa grandis, Stipa kyrilovii, Artemisia frigida* and various cold and salt tolerant shrubs and subshrubs dominate. The botanical composition of alpine meadows depends to a large extent on the grazing pressure and seasonal management regime [1,10,18]. For example, Table 2 shows the list of common forage species and their ranking (based on abundance) in grasslands grazed under contrasting management regimes.

There was more sustained grazing pressure on the SH areas than on the MH sites because of the effect of trampling [18,19] and non-selective grazing. We measured the projected foliage cover to act as an indicator of grazing pressure along with biomass per m^2 and species abundance (Table 2). In the SH sites *Ligularia virgaurea* (9.3% ± 10% projected foliage cove [PFC]), *Anemone rivularis* var. *flore-minore* (5.4 ± 3.9 PFC) and *Taraxacum mongolicum* (5.3% ± 3.4% PFC) are regarded as indicators of land degradation because they are toxic or unpalatable plants that exploit gaps exposed by grazing pressure. Poisonous plants were observed to be more common on SH land. This suggests that the SH management regime may be causing further serious degradation and that productive forage species may be on the decline [20–22]. Field work to substantiate this is underway.

Apart from the actual management regime (where, when and the duration of grazing at any specific location) there is the grazing pressure or stocking rate. As Table 3 shows there is often a wide disparity between the actual and the recommended stocking rates. Because of the high altitude and short growing season, livestock are moved from the higher grasslands in summer and autumn to the cool season grasslands at lower altitudes (often referred to as winter pastures. herders typically have a 'winter residence' where livestock, principally yaks, can overwinter in the open [22,23].

Table 2. The most commonly found forage species in grassland grazed under two different systems. **SH** refers to an area where a single household makes all the decisions about when, where and how long to graze a specific site and **MH** refers to an area under a community-based management scheme where cooperating households follow a plan, including scheduled bans to allow rest rotation grazing.

Plant Species	SH g/m^2 ± SE	Rank # SH	MH g/m^2 ± SE	Rank MH
Anenome rivularis var. *flore-minore* (F)	5.9 (±3.6)	8	–	–
Elymus nutans (G)	13 (±16)	6		
Festuca sinensis (G)	14 (±8)	4	25 (±11)	2
Guildenstaedia diversifolia (L)			22 (±14)	4
Kobresia capillifolia (S)	22 (±21)	1	24 (±20)	3
Kobresia humilis (S)	4.1 (±1.7)	10 (eq.)		
Lancea tibetica (F)			19.3 (±10.7)	6
Poa pratensis (G)			20 (±12)	5
Potentilla fragarioides (F)	15.6 (±11)	3		8
Ligularia virguarea (F)	11.9 (±10.7)	7		
Ranunculus angustifolia var. *capillarus* (F)			12.9 (±9)	10
Scirpus distigmaticus (G)	19 (±17)	2	26 (±15)	1
Taraxacum mongolicum (F)	5.3 (±3.4)	10 (eq.)		
Thalicrum alpinum (F)	13 (±8.1)	5		

G—grass; F—forb; S—sedge; L—legume (highly preferred by livestock). # Rank is based on species abundance (measured as biomass and expressed as g/m^2). Source: Squires and Feng unpublished. Data collected from numerous sites (>70) within a long-term trial designed by Lanzhou University in Maduo county to compare two grazing regimes. The data show a high degree of variability but are enough to show the major trends.

Table 3. Contrast between practical stocking rate (S/R) and theoretical stocking rate of grassland in five counties in Qinghai (×10^4 sheep unit) *.

Area/County	Maduo	Maqin	Dari	Gande	Qumalai
Livestock number #	84.5	138.3	114.2	105.7	142.5
Theoretical S/R In cool season	66.7	99.2	124.1	29.6	70.1
Theoretical S/R In warm season	304.9	196.6	267.5	165.6	98.5
The ratio of duration of time spent between the cool and warm season * area	Around 2:1				

Note: # One sheep unit is the equivalent of a 40-kg sheep. This enables different types and sizes of livestock to be expressed as a common unit using approved conversion ratios for goats, camels, yaks, horses, etc. * Cool and warm are relative terms. The cool season areas are at lower elevations and herds and their herders over-winter there. This is despite subzero temperatures for months on end. Data source: Feng unpublished, based on advice from local animal husbandry and grassland management bureaus in each county and a synthesis of published studies.

2.3. The Important Role of Alpine Meadows

Alpine grasslands, especially alpine meadows, make a huge contribution to the ecology and biodiversity of plants and animals (including microfauna) of the QTP and by extension, to the whole of China and even to countries to the south-east and those further west (Bangladesh, India) that depend on rivers that arise on the QTP (Figure 3b).

Alpine meadows, because of their sod-forming habit and their role in soil protection [23,24] are particularly important in the headwater regions on the QTP. The largest and most important of these is the so-called Three Rivers Regions (Sanjiangyuan). It covers an area of 152,300 km^2 and lies at an altitude between 3500 and 4600 m ASL. To protect the head waters of many important rivers (including five that are international rivers, there are plans to remove more than 200,000 herders and their livestock to new resettlement areas outside the Sanjiangyuan region. On one hand, there is the laudable attempt to conserve biodiversity and habitat [12,24] that is currently under threat from too many people and too many livestock competing for water and forage. The success or otherwise of the resettlement is the subject of considerable debate [25–30]. While on the other hand there is the sociocultural devastation as people, who until recently were semi nomadic and who (with their livestock) sought forage and water across the high plateau that is their spiritual home. Well-intentioned land management efforts, such as fencing, cause land fragmentation [31]. Restrictions on herder mobility cause concentration of livestock in critical 'refuge areas'. The changes in the length of the growing season (earlier 'green-up' and later senescence in the autumn) encourage extension of the grazing on 'summer pastures' to the detriment of native ungulates (and other fauna) that overwinter there [13,24,25]. See discussion on phenology in Section 4 below.

The landscape is characterized by expansive and flat grassland interspersed with both rounded and high mountains whose surface was weathered to such a degree that grassland is distributed everywhere except at the top above the snowline. The vegetation is dominated by grasses or 'grass like' species, for example, *Kobresia* spp.—a sedge.

According to [5] these alpine grasslands are predominantly in 'intact' natural condition in many areas. These grasslands though, are under threat of increasingly strong human activities such as overgrazing, cultivation, collection of fuel and medicinal plants, open pit mining, infrastructure development (roads, rail, transmission lines, rights of way, fencing) land fragmentation, as well as climate change [9,10,23]. Dominated by Lamaist pastoral cultures for millennia (Lamaist peoples are pastoralists grazing yaks, Tibetan sheep, goats, Bactrian camels and horses throughout the QTP rangelands. Most the local people practice Lamaism, a pastoral religion that combines Shamanism and Buddhism with a sprinkling of Dao (Taoist) symbolism). The rangelands in many parts are now almost completely devoid of woody plant species (the QTP is mostly naturally treeless except in the south-eastern river valleys). As a result, the degraded rangelands are subject to accelerated runoff, severe desertification and unpredictable climatic extremes due to widespread disruptions to heat–water cycle.

As Brierley [32] said " ... our present understanding ... is insufficient to ensure sustainable approaches to environmental management [on the QTP]. A key requirement for this to be achieved is effective understanding of the diversity of grassland types, their character/behavior, their pattern/distribution, their evolutionary trajectory, their linkages to other elements of landscapes, their vulnerability to disturbance (whether climate change or land use changes) and the ecological values of grassland ecosystems that we seek to protect". To this may be added changes in the policy environment as it relates to use rights, land tenure and changes in rural sociology of land users [33–37].

Harris [33] identified and assessed twelve factors implicated in grassland degradation. Half of these are biophysical in nature; the remaining six are human factors all influenced by wider socioeconomic processes, pointing to a need for better understanding of socio-ecological systems in the QTP as a basis for developing policy initiatives. In particular it is critical to investigate the rationale underlying 'traditional' grazing practices and ascertain the role of practices grazing (both good and bad) in grassland degradation. The common assumption that overgrazing is a key factor is questioned by some authors [12,24,25,33] yet this influences current policies such as sedentarization and new fencing regimes that appear to produce mixed ecological outcomes [36,37].

Until relatively recently, the long-term ecological implications of privatizing grassland and reducing spatial movement (including transhumance) had received little analysis despite early work by [24,36,38]. Arguments are mounting that flexible grazing practices are essential if there is to be sustainable grassland management [24,25,27,31].

3. Environmental Changes on Grasslands of the QTP under Global Change

An analysis of impacts of climate change on the grassland ecosystems of the Qinghai–Tibet Plateau (QTP) shows that the grasslands are sensitive to climate change and human disturbance [20,24]. This huge geographic region is also known as the "Third Pole". The ecosystems on the QTP play important roles in servicing the ecological environment and regional (SE Asian) socio-economic development and in the maintenance of biodiversity of plants and animals [39,40].

The grasslands of the QTP are the core of the water cycle, and their health directly affects the flux of water and water conservation capacity. In terms of glacier water, the glacier area is about 49,873 km^2 and ice reserves about 4561 km^3. The frozen soil area is 2.7×10^6 km^2, accounting for 46% of the total Qinghai–Tibetan Plateau land area. The groundwater in the QTP is the source of runoff water of many surrounding low altitude areas. The lake census showed that there are 32,843 lakes with a total area of 43,151 km^2, which makes up 1.4% of the total area of the Plateau. Around 96% of all lakes are small, with an area <1 km^2, while the 1204 large lakes (>1 km^2) account for a small percentage of the total area of lakes [41].

Climate change has already produced visible adverse effects on agriculture and livestock-raising sectors, manifested by increased instability in agricultural production, severe damage to crops and livestock breeding caused by drought and high temperature in some parts of the QTP. Moreover, under the background of climate change, herders have to adapt to the changes in terms of their production pattern as well as livelihood style. Therefore, understanding the effect of climate change on grassland production and response of herders to climate change is vital to improve grassland management on the QTP [1,42–44].

From 1951 to 2009, the ground surface temperature increased on average by 1.38 °C in China, the warming rates is 0.23 °C per 10 years, which is similar to the level of global warming. The warming on the QTP was greater than in other regions, especially in winter and spring. Most regions of QTP, including Xizang (Tibet) and Qinghai Province, have a greater warming rate compared to other regions. The climate in west and north QTP has changed from warmer-wetter to warmer-dryer, however, the climate in the east and south of the QTP has changed from warmer-dryer to warmer-wetter [38,41–43]. The precipitation on the QTP was generally rising in the 30 years ending 2012 (12.4 mm per 10 years) [42,45,46], in particular in the north of the QTP. The depth snow and the number of days with snow cover on the QTP increased from 1961 to 1990 and decreased from 1991 to 2005 [37,44,45].

IPCC [44] reported that the occurrence of extreme events is rising due to climate change. The snow disaster was also rising in recent years. Snow disaster or *dzud* (in Mongolian language and *kengschi* in Tibetan) is where deep snow, severe cold or other conditions make forage inaccessible or unavailable and leads to high livestock mortality. *Dzud* is a recurring natural event that limits the growth of livestock populations on the QTP, but it also causes loss of human life and livelihoods. In the *dzud* events of 1999–2002 and 2009–2010 the QTP lost 30% and 20% of the entire livestock flocks/herds, respectively, creating significant challenges for herders and rural communities who service the herders and their families. A herder's sole source of income can be wiped out by a *dzud*. On the QTP there are over one million herders and their families. In the 60 years ending in 2012, the snow disaster occurred 30 times in Qinghai Province, among them, severe snow disaster occurred 12 times and extreme snow disaster six times. From 1949 to 2002, the snow disaster occurred 80 times in Xizang (Tibet) and snow disasters that caused livestock mortalities in excess of two million head have occurred seven times [45]. The impact on people and their livelihoods is enormous [9,45,46].

4. Science and the Pastoral Management Interface on the QTP

The ongoing debate over spring vegetation phenological trends on the Tibetan Plateau, and the scientific disagreement over the role of climate change in driving these changes, illustrates the limitations of current scientific observations and knowledge of climate change and its impacts due to system complexity, heterogeneity and data scarcity (Klein et al. [47]. For example, NDVI-based phenology studies of Tibet show opposing trends (Chen et al. [48]; Shen, [49]; Shen et al. [50,51]; Wang and Chang, [52]; Yu et al., [53]; Zhang et al. [54]. Tibetan pastoralists perceived delays in the start of summer over multi-decade timescales, despite the shorter-term reversal of this trend revealed by NDVI measurements. The longer term NDVI trend toward delayed green-up corresponds with herders' local observations of shifts in seasonality, indicating that their long-term, qualitative perceptions support quantitative patterns deemed non-significant by Western scientific methods. In our experience, respondents who were engaged in herding as their primary livelihood consistently observed summers starting later, whereas those mostly engaged in non-herding livelihood activities were split, with those in the higher latitudes (areas with generally more mountainous terrain and less vegetative production) observing delayed summers, and those in the lower latitudes (areas with generally less mountainous terrain and higher vegetative production) split between observing no change and delayed summers. Klein et al. [47], and our own observations, suggest that policy interventions driven by scientific data derived from NVDI may run counter to pastoralists perceptions and ignore vital shifts in the availability of forage and water.

5. Conclusions

The alpine grasslands of the QTP have undergone major change in the past 70 years or so. Anthropogenic pressure has increased as populations of people and their livestock have more than doubled. In the past the alpine grasslands (the principal grazing lands) were under control of the local Buddhist temple, herders at that time followed directives about where and when to graze and mobility of livestock ensured that the herders had maximum access to forage and water [36,38,39]. In more recent decades there was phase of collectivization when government owned all the livestock and the land. Later in the 1980s, under the household contract responsibility system, ownership of livestock and grazing use rights to specified tracts of grassland passed to households. This created an anomaly in that livestock were privately owned and grazed on public land under a system of common use grazing [23]. Constraints on livestock inventories were very lax and livestock populations soared. Of late the government has tried to foster their national program to "Return grazing lands to grassland". Livestock numbers were restricted under this scheme and local animal husbandry and grassland monitoring bureaus have imposed grazing bans and erected fencing to regulate when and where livestock can graze. In situations where land degradation was severe, resettlement of herders (see above) and reduction of herd numbers has been implemented. About 220,000 herder families are to be moved from the Three Rivers region where the headwaters of the Yellow, Yangtze and Mekong Rivers are located. The grassland dynamics have changed a lot and now there is the increasing impact of greater climate variability as part of climate change [1,42,46].

The QTP is a vast and complex region characterized by harsh environmental conditions and mired in several policy gridlocks as government tries to meet its obligations under the various UN Conventions on biodiversity, climate and desertification as well as on human rights. There is a desire to 'close the gap' between the urbanized eastern part of China and the generally less well developed western regions (including the QTP) while maintaining social harmony as people are re-located in a bid to modernize the rural areas and deliver a wide range of social services (schools, clinics, pension schemes) and better infrastructure such as rural electrification, roads and trains, etc. Land use on the grasslands that have for millennia supported the hardy semi-nomadic herders, are now being re-organized and re-allocated. Despite these changes there is a vast area of alpine grassland that is a sink for carbon dioxide and habitat for a range of emblematic fauna including predators such as the snow leopard and numerous birds, rodent and fish that cannot be found elsewhere on the globe.

Author Contributions: Conceptualization, V.R.S. and H.F.; methodology, H.F.; validation, H.F. and V.R.S.; investigation, H.F.; resources, H.F.; data curation, H.F.; writing—original draft preparation, V.R.S.; writing—review and editing, V.R.S.; supervision, H.F.; project administration, H.F.; funding acquisition, H.F. All authors have read and agree to the published version of the manuscript.

Funding: One of us (HYF) was in receipt of a grant from the National Social Science Fund of China Grant number 17BMZ106 to investigate and report on *"Grassland degradation and herders' sustainable livelihood on the Qinghai Tibetan Plateau"*.

Acknowledgments: We are grateful to the many researchers, commentators and scholars whose work has inspired our own. We have drawn heavily on their wisdom, insights and writings to bring what we hope will be a synthesis. We are thankful for the support we received from government officials and local leaders as well as the hospitality shown by Tibetan herders in Qinghai and beyond. We are grateful to Mahesh Gaur for re-drawing and/or enhancing the Figures.

Conflicts of Interest: The authors declare no conflict of interest.

References

1. Feng, H.; Squires, V.R. (Eds.) *Climate Variability Impacts on Land Use and Livelihoods in Drylands*; Springer International Publishing: New York, NY, USA, 2018; pp. 91–112.
2. Miehe, G.; Schleuss, P.-M.; Seeber, E.; Babel, W.; Biermann, T.; Braendle, M.; Chen, F.; Coners, H.; Foken, T.; Gerken, T.; et al. The Kobresia pygmaea ecosystem of the Tibetan highlands—Origin, functioning and degradation of the world's largest pastoral alpine ecosystem. *Sci. Total Environ.* **2019**, *648*, 754–771. [CrossRef] [PubMed]
3. Dinesman, L.G.; Kiseleva, N.K.; Knyazev, A.V. *History of the Steppe Ecosystems of the Mongolian People's Republic*; Nauka: Moscow, Russia, 1989.
4. Pfeiffer, M.; Dalamsuren, C.H.; Jäschke, Y.; Wesche, K. Grasslands of China and Mongolia: Spatial Extent, Land Use and Conservation. In *Grasslands of the World: Diversity, Management and Conservation*; Squires, V.R., Dengler, J., Feng, H., Hua, L., Eds.; CRC Press: Boca Raton, FL, USA, 2018; pp. 168–196.
5. Li, F.Y.; Jäschke, Y.; Guo, K.; Wesche, K. *Grasslands of China*; Elsevier: Amsterdam, The Netherlands, 2019.
6. Qiu, Q.; Wang, L.; Wang, K.; Yang, Y.; Ma, T.; Wang, Z.; Zhang, X.; Ni, Z.; Hou, F.; Long, R.; et al. Yak whole-genome resequencing reveals domestication signatures and prehistoric population expansions. *Nat. Commun.* **2015**, *6*, 10283. [CrossRef] [PubMed]
7. Bartholome, E.; Belward, A.S. GLC2000: A new approach to global land cover mapping from Earth observation data. *Int. J. Remote Sens.* **2005**, *26*, 1959–1977. [CrossRef]
8. Sun, H. *Ecosystems of China*; Science Press: Beijing, China, 2005.
9. Gao, Q.; Li, Y.; Xu, H.-M.; Wan, Y.-F.; Jiangcun, W.-Z. Adaptation strategies of climate variability impacts on alpine grassland ecosystems in Tibetan Plateau. *Mitig. Adapt. Strat. Glob. Chang.* **2012**, *19*, 199–209. [CrossRef]
10. Wang, J.; Zhang, J.; Zhou, X. The Qinghai-Tibet Plateau Alpine Vegetation Region. In *Vegetation of China*; Wu, Z., Ed.; Beijing Science Press: Beijing, China, 1980; pp. 1022–1071.
11. Global Carex Group. Making Carex monophyletic (Cyperaceae, tribe Cariceae): A new broader circumscription. *Bot. J. Linn. Soc.* **2015**, *179*, 1–42. [CrossRef]
12. Gu, A. Biodiversity of Tibet's rangeland resources and their protection. In *Tibet's Biodiversity: Conservation and Management*; Lu, Z., Springer, J., Eds.; China Forestry Press: Beijing, China, 2000; pp. 94–99.
13. Wu, C.; Wang, H. *Physical Geography of China: Phytogeography (I)*; Science Press: Beijing, China, 1983.
14. Ellenberg, H.; Mueller-Dombois, D. *Tentative Physiognomic-Ecological Classification of Plant Formations of the Earth*; Ber. Geobot. Inst. Rubel Zurich: Zurich, Switzerland, 1965; Volume 37, pp. 21–55.
15. Ellenberg, H.; Mueller-Dombois, D. Aims and Methods of Vegetation Ecology. *Geogr. Rev.* **2013**, *66*, 114–116.
16. Luo, Q.; Ni, L.M.; He, B.; Wu, H.; Xue, W. MEADOWS: Modeling, emulation, and analysis of data of wireless sensor networks. In Proceedings of the Proceeedings of the 1st International Workshop on Data Management for Sensor Networks: In Conjunction with VLDB 2004, Toronto, ON, Canada, 30 August 2004; pp. 628–665. [CrossRef]
17. Kang, L.; Han, X.; Zhang, Z.; Sun, O.J. Grassland ecosystems in China: Review of current knowledge and research advancement. *Philos. Trans. R. Soc. B Boil. Sci.* **2007**, *362*, 997–1008. [CrossRef]
18. Ao, M.; Ito, M.; Ito, K.; Yun, J.F.; Miura, R.; Tominaga, T. Floristic compositions of Inner Mongolian grasslands under different land-use conditions. *Grassl. Sci.* **2008**, *54*, 173–178. [CrossRef]
19. Xie, T.P.; Du, G.Z.; Zhang, G.F. Effects of inflorescence position on seed production and seedling establishment in Ligularia virgaurea. *Chin. J. Plant Ecol.* **2010**, *34*, 418–426.
20. Yan, Z.L.; Wu, N.; Dorji, Y.; Jia, R. A review of rangeland and privatization and its implication in the Tibetan Plateau, China. *Nomad. Peoples* **2005**, *9*, 31–51.
21. Shang, Z.H.; Dong, Q.; Degen, A.; Long, R.J. Ecological restoration on Qinghai-Tibetan Plateau: Problems, Strategies and Prospects. In *Ecological Restoration: Global Challenges, Social Aspects and Environmental Benefits*; Squires, V.R., Ed.; Nova Science Publishers: New York, NY, USA, 2016; pp. 151–176.
22. Shang, Z.; Gibb, M.J.; Leiber, F.; Ismail, M.; Ding, L.M.; Guo, X.S.; Long, R.J. The sustainable development of grassland-livestock systems on the Tibetan plateau: Problems, strategies and prospects. *Rangel. J.* **2014**, *36*, 267–296. [CrossRef]
23. Li, M.; Zhang, X.; He, Y.; Niu, B.; Wu, J. Assessment of the vulnerability of alpine grasslands on the Qinghai-Tibetan Plateau. *PeerJ* **2020**, *8*, e8513. [CrossRef] [PubMed]

24. Niu, Y.; Zhou, J.; Yang, S.; Chu, B.; Ma, S.; Zhu, H.; Hua, L. The effects of topographical factors on the distribution of plant communities in a mountain meadow on the Tibetan Plateau as a foundation for target-oriented management. *Ecol. Indic.* **2019**, *106*. [CrossRef]
25. Du, F. Ecological Resettlement of Tibetan Herders in the Sanjiangyuan: A Case Study in Maduo County of Qinghai. *Nomad. Peoples* **2012**, *16*, 116–133. [CrossRef]
26. Foggin, J.M. Rethinking "Ecological migration" and the value of cultural community: A response to Wang. *J. Hum. Environ.* **2011**, *40*, 100–101. [CrossRef]
27. Zheng, R. Analysis on resettlement in ecologically vulnerable areas of western China: Resettlement policy and practices aiming at ecological environmental protection and poverty alleviation. *Yangtze River* **2011**, *42*, 93–99.
28. Tashi, G.; Foggin, J.M. Resettlement in Development and Progress? Eight years on. Review of emerging Social and Development impacts on "Ecological resettlement Project" in Tibet Autonomous Region, China. *Nomad. Peoples* **2012**, *16*, 134–151. [CrossRef]
29. Tsering, D. The theoretical relations between the poverty-oriented resettlement and environmental migration. *Tibet. Stud.* **2014**, *5*, 45.
30. Wang, X. Studies on ecological refugees in the source region of the Yellow River, Lancang River and Yangtze River. *J. Qinghai Nationalities Univ.* **2006**, *17*, 23–31.
31. Squires, V.R.; Hua, L.M. Land fragmentation: A scourge in China's pastoral areas. *Livest. Res. Rural Dev.* **2017**, *29*, 1–16.
32. Brierley, G. Framing wetlands in their landscape context: Scientific and Management Implications. In *Wetland Types and Evolution and Rehabilitation in the Sanjiangyuan Region*; Chen, G., Li, X., Gao, J., Brierley, G., Eds.; Qinghai People's Publishing House: Xining, China, 2010; pp. 9–16.
33. Harris, R.B. Rangeland degradation on the Qinghai Tibet Plateau: A review of the evidence of its magnitude and causes. *J. Arid. Environ.* **2010**, *74*, 1–12. [CrossRef]
34. Miller, D.J. Tough Times for Tibetan Nomads in Western China: Snowstorms, Settling down, Fences and the Demise of Traditional Nomadic Pastoralism. *Nomad Peoples* **2000**, *4*, 83–109. [CrossRef]
35. Foggin, M. Depopulating the Tibetan Grasslands. *Mt. Res. Dev.* **2008**, *28*, 26–31. [CrossRef]
36. Yeh, E.T. Greening western China: A critical view. *Geoforum* **2009**, *40*, 884–894. [CrossRef]
37. Sheehy, D.P.; Miller, D.; Johnson, D.A. Transformation of traditional pastoral livestock systems on the Tibetan steppe. *Sécheresse* **2006**, *17*, 142–151.
38. Zhang, G.; Yao, T.; Xie, H.; Kang, S.; Lei, Y. Increased Mass over the Qinghai-Tibet Plateau: From lakes or glaciers? *Geophys. Res. Lett.* **2013**, *40*, 2125–2130. [CrossRef]
39. Zhao, C.; Squires, V. Biodiversity of Plants and Animals in Mountain Ecosystems. In *Towards Sustainable Use of Rangelands in North-West China*; Springer Science and Business Media LLC: Berlin/Heidelberg, Germany, 2010; pp. 101–125.
40. Squires, V.R.; Safarov, N.M. Diversity of plants and animals in mountain systems in Tajikistan. *J. Rangel. Sci.* **2013**, *4*, 43–61.
41. Xu, W.; Liu, X. Response of vegetation in the Qinghai-Tibetan Plateau to global warming. *Chin. Geogr. Sci.* **2007**, *17*, 151–159. [CrossRef]
42. Liang, Y.; Ganjurjav, W.N.; Zhang, B.; Qing-zhu, G.; Luo-bu, D.; Zhuo-ma, X.; Yu-zhen, B. A review on effect of climate change on grassland ecosystems in China. *J. Agric. Sci. Technol.* **2014**, *16*, 1–8.
43. Zhao, Y.; Ding, Y.; Hou, X.; Li, F.Y.; Han, W.; Yun, X. Effects of temperature and grazing on soil organic carbon storage in grasslands along the Eurasian steppe eastern transect. *PLoS ONE* **2017**, *12*, e0186980. [CrossRef] [PubMed]
44. IPCC. *AR5 Synthesis Report: Climate Change. The Synthesis Report of the IPCC Fifth Assessment Report*; Intergovernmental Panel: Geneva, Switzerland, 2014.
45. Liu, F.; Mao, X.; Zhang, Y.; Chen, Q.; Liu, P.; Zhao, Z. Risk analysis of snow disaster in the pastoral areas of the Qinghai-Tibet Plateau. *J. Geogr. Sci.* **2014**, *24*, 411–426. [CrossRef]
46. Hua, L.M.; Niu, Y.; Squires, V.R. Climate change on grassland regions and its impact on grassland-based livelihoods in China. In *Grasslands of the World: Diversity, Management and Conservation*; Squires, V.R., Dengler, J., Feng, H., Hua, L., Eds.; CRC Press: Boca Raton, FL, USA, 2018; pp. 355–368.

47. Klein, J.A.; Hopping, K.A.; Yeh, E.T.; Nyima, Y.; Boone, R.B.; Galvin, K.A. Unexpected climate impacts on the Tibetan Plateau: Local and scientific knowledge in findings of delayed summer. *Glob. Environ. Chang.* **2014**, *28*, 141–152. [CrossRef]
48. Chen, H.; Zhu, Q.; Wu, N.; Wang, Y.; Peng, C.-H. Delayed spring phenology on the Tibetan Plateau may also be attributable to other factors than winter and spring warming. *Proc. Natl. Acad. Sci. USA* **2011**, *108*, E93. [CrossRef]
49. Shen, M.; Tang, Y.; Chen, J.; Zhu, X.; Zheng, Y. Influences of temperature and precipitation before the growing season on spring phenology in grasslands of the central and eastern Qinghai-Tibetan Plateau. *Agric. For. Meteorol.* **2011**, *151*, 1711–1722. [CrossRef]
50. Shen, M. Spring phenology was not consistently related to winter warming on the Tibetan Plateau. *Proc. Natl. Acad. Sci. USA* **2011**, *108*, E91–E92. [CrossRef]
51. Shen, M.; Sun, Z.; Wang, S.; Zhang, G.; Kong, W.; Chen, A.; Piao, S. No evidence of continuously advanced green-up dates in the Tibetan Plateau over the last decade. *Proc. Natl. Acad. Sci. USA* **2013**, *110*, E2329. [CrossRef]
52. Wang, T.; Peng, S.; Lin, X.; Chang, J. Declining snow cover may affect spring phenological trend on the Tibetan Plateau. *Proc. Natl. Acad. Sci. USA* **2013**, *110*, E2854–E2855. [CrossRef]
53. Yu, H.; Luedeling, E.; Xu, J. Winter and spring warming result in delayed spring phenology on the Tibetan Plateau. *Proc. Natl. Acad. Sci. USA* **2010**, *107*, 22151–22156. [CrossRef]
54. Zhang, G.; Zhang, Y.; Dong, J.; Xiao, X. Green-up dates in the Tibetan plateau have continuously advanced from 1982 to 2011. *Proc. Natl. Acad. Sci. USA* **2013**, *110*, 4309–4314. [CrossRef]

MDPI
St. Alban-Anlage 66
4052 Basel
Switzerland
Tel. +41 61 683 77 34
Fax +41 61 302 89 18
www.mdpi.com

Applied Sciences Editorial Office
E-mail: applsci@mdpi.com
www.mdpi.com/journal/applsci

www.ingramcontent.com/pod-product-compliance
Lightning Source LLC
LaVergne TN
LVHW070616170726
843515LV00004B/98